THE IMPACT OF SOIL EROSION IN THE UPPER BLUE NILE ON DOWNSTREAM RESERVOIR SEDIMENTATION

THE IMPACT OF SOIL EROSION IN THE UPPER BLUE NILE ON DOWNSTREAM RESERVOIR SEDIMENTATION

DISSERTATION
Submitted in fulfilment of the requirements of
the Board for Doctorates of Delft University of Technology
and of the Academic Board of the UNESCO-IHE Institute for Water Education
for the Degree of DOCTOR
to be defended in public
on Tuesday October 28th, 2014, at 10:00 hours
in Delft, The Netherlands

by

YASIR SALIH AHMED ALI

born in Wad Medani, Sudan
BSc Civil Engineering, University of Khartoum, Khartoum, Sudan
MSc Water Management, Water Management and Irrigation Institute, University of Gezira, Wad Medani, Sudan
MSc Water Science and Engineering 'Hydraulic Engineering and River Basin Development', UNESCO-IHE, Delft, the Netherlands

CRC Press is an imprint of the
Taylor & Francis Group, an **informa** business
A BALKEMA BOOK

This dissertation has been approved by the promotors:
Prof.dr.ir. A.E. Mynett
Prof.dr. N.G. Wright

Composition of the Doctoral Committee:

Chairman:	Rector Magnificus, Delft University of Technology
Vice-chairman:	Rector, UNESCO-IHE
Prof.dr.ir. A.E. Mynett	UNESCO-IHE/ Delft University of Technology, promotor
Prof.dr. N.G. Wright	UNESCO-IHE/ University of Leeds, promotor
Prof.dr.ir. N.C. van de Giesen	Delft University of Technology
Dr.ir. A. Crosato	UNESCO-IHE/ Delft University of Technology
Dr.ir. A.J.F. Hoitink	Wageningen University
Prof.dr. Y.A. Mohamed	HRC, Sudan/UNESCO-IHE
Prof.dr.ir. W.Uijttewaal	Delft University of Technology, reserve member

First issued in hardback 2018

CRC Press/Balkema is an imprint of the Taylor & Francis Group, an informa business

Published by:

CRC Press/Balkema
PO Box 11320, 2301 EH Leiden, The Netherlands
e-mail: Pub.NL@taylorandfrancis.com
www.crcpress.com – www.taylorandfrancis.co.uk

ISBN 13: 978-1-138-37326-6 (hbk)
ISBN 13: 978-1-138-02742-8 (pbk)

To the memory of my lovely sister, Nada

SUMMARY

Population growth in the upper Blue Nile Basin led to fast land-use changes from natural forest to agricultural land, which resulted in speeding up the soil erosion processes. Soil erosion undesirably reduces soil fertility and hence the agricultural productivity upstream. Eroded sediment is transported to the lower Blue Nile Basin, where sedimentation occurs at many locations. In the reservoirs, sedimentation leads to serious reduction in storage capacity, causing hydropower generation problems and negative impacts on the socio-economic, environmental and ecological system. The sediment settling inside irrigation canals leads to water shortage and management difficulties. Sediment deposition in the main channel of the river raises the bed level and enhances flood risks.

The Blue Nile River Basin is currently experiencing new developments, both in Ethiopia and Sudan. The Grand Ethiopia Renaissance Dam (GERD) is under construction about 30 km upstream of the Ethiopian-Sudanese border. Recently, the Roseires Dam located 110 km downstream the Ethiopian-Sudanese border has been heightened by 10 m, increasing the storage capacity of the reservoir by additional 3700 million m³. Some dams are planned in Ethiopia for hydropower production. These developments will strongly affect the water resources and sediment deposition in the lower Blue Nile Basin.

Sedimentation in the new reservoirs and in irrigation canals will depend on the operation of these dams, but the only effective solution to reduce the sedimentation problems is reducing the sediment input. This can be achieved by means of erosion control practices in the upper basin. For this, given the vastness of the upper basin, it is important to identify the areas where the largest amounts of sediment are produced.

The main objective of this research is to identify these areas and quantify the amounts of sediment involved. This research investigates also the effects of the new developments on the sediment processes. Several knowledge gaps have been filled in by this research in order to fulfill the goals. Missing information comprised:

- ➤ Bathymetric and morphological data including river cross-sections, as well as river bed and bank material along the main river and tributaries.
- ➤ Hydrological data including flow discharges and sediment concentration.
- ➤ Water and sediment transport distribution along the river network at all flow conditions.
- ➤ History of sedimentation (including amounts) in the basin.
- ➤ Origin of the deposited sediment in the lower basin.
- ➤ Relation between land-use changes and sediment yield in the sub basins providing the largest amounts of sediment deposition in the lower basin.

Extensive field surveys were conducted both in Ethiopia and Sudan as part of this study to bridge the knowledge gaps. The bed topography was measured at 26 cross sections along the Blue Nile River and major tributaries using an eco-sounder in Ethiopia and an Acoustic Doppler Current Profiler in Sudan. Soil samples were taken from the areas affected by erosion in the upper basin and from the bed and banks of the rivers. Suspended sediment samples were collected at several locations along the Blue Nile River and its tributaries, both in Ethiopia and Sudan. In addition, suspended sediment concentration was sampled on a daily basis near the Ethiopian-Sudanese border during the flood season for 4 years. The collected samples were analyzed at Addis Ababa University laboratory, the Hydraulics Research Station laboratory in Wad Medani and the Technical University of Delft.

Annual flow discharge and sediment load balances were obtained by integrating the available and newly measured flow discharges, suspended sediment concentration and the results of numerical models using the Soil and Water Assessment Tool software (SWAT). The yearly sediment balances were estimated at several locations along the main river and the tributaries. Three regression approaches were used to determine the sediment loads from the rating curves derived from the measured data. These were developed using the linear and non-linear log-log regressions, while the statistical bias correction factor was used to improve the linear regression results.

The water distribution along the entire river system was assessed in order to quantify the availability of the water resource at all seasons and flow conditions. A one-dimensional hydrodynamic model of the entire river network was developed including all known water uses for irrigation, as well as all major hydraulic structures and their operation rules using the Sobek software. The model was further used to study sediment transport via integration with the water quality module of the Delft3D software. This integrated model (Sobek Rural and Delft3D Delwaq) allowed simulating the morphololological processes along the Blue Nile River, from Lake Tana to Roseires Dam. The model was calibrated and validated based on the Roseires Reservoir historical bathymetric surveys, and the sediment concentrations measured at the Ethiopian-Sudanese border. The model was then run to predict the impact of Roseires Dam heightening and the construction of the Grand Ethiopian Renaissance Dam on sedimentation rates.

The history of sedimentation inside Roseires Reservoir, the first sediment trap along the Blue Nile River, was studied by combining historical bathymetric data with the results of a quasi 3D morphodynamic model including vertical sorting (based on the Delft3D software). Selective sedimentation creates soil stratification inside the reservoir allowing for the recognition of specific wet or dry years. The most promising coring locations from where soil samples could be collected were identified by analyzing the results of the model, since the model allowed identifying the areas that were neither subject to net erosion nor to bar migration during the life span of the reservoir. A second measuring campaign took place at these locations to analyze the sediment deposited in the reservoir

The origin of the sediment deposited in the reservoir was indentified based on the mineral characteristics of the material. X-Ray Diffraction laboratory analyses allowed assessing the mineral content in the sediment samples that were collected during the field campaigns from the eroded areas in the upper basins and from the deposited soil layers inside Roseires Reservoir. The integration of the results of the X-Ray Diffraction with a cluster analysis allowed identifying the source of the sediments deposited inside Roseires Reservoir. The results showed that the sub basins of Jemma, Didessa and South Gojam are the main sediment source areas. The implementation of erosion control practices can therefore start from these sub basins. The land use changes occurred in these sub basins in the last 40 years show that natural forest, woodland, wooded grassland and grassland decreased from more than 70% to less than 25% of the surface area. Instead, the cultivated area increased from 30% to more than 70% of the total surface area.

Finally, model results showed that the annual sediment deposition inside the Renaissance Dam (under construction) will vary with time, with maximum and minimum values of 45 and 17 million m³/year, respectively, and an average deposition rate of 27 million m³/year. The average deposition rate in Roseires Reservoir after heightening and Renaissance Dam construction was found to be 2 million m³/year. This means that the annual sedimentation rates inside Roseires Reservoir will be reduced by more than 50% if compared with the present situation. These results are affected by a high level of uncertainty, but it can be assumed that the trend and order of magnitude are represented reasonably well by the calibrated and validated model.

ACKNOWLEDGEMENTS

The study was carried out as a project within a larger research program called "In search of sustainable catchments and basin-wide solidarities in the Blue Nile River Basin", funded by the Foundation for the Advancement of Tropical Research (WOTRO) of the Netherlands Organization for Scientific Research (NWO). I would like to thank WOTRO and UNESCO-IHE for financial support to carry out this doctoral study. I would like to express my sincere appreciation to all project leaders that directly or indirectly stimulated me to conduct this research. Among them the project leader Prof. Pieter van der Zaag, Prof. Stefan Uhlenbrook, Dr. Belay Semane, Prof. Sief ElDin Hamad Abdalla and Prof. Abdeen Salih.

I would like to express my deep and sincere gratitude to my promotor Prof. Nigel Wright for giving me the chance to start this PhD research and supervising me systematically and tirelessly through both my MSc and PhD research during the past seven years.

I would like to extend my deepest gratitude to my promotor Prof. Arthur Mynett for his follow up of my progress, good inspiration, scientific advice and support. From him I have learnt a lot of Knowledge when I was an Msc student.

I owe a great deal of appreciation for this thesis to my supervisor Dr. Crosato, she has been sharing her vision and wisdom, trusting my capability in doing research, giving me freedom to develop new ideas and allowing me to attend quite a number of international conferences as part of this research. Thank you very much for your continued support; without you, this study would not have been finished successfully.

I am highly thankful to Dr Yasir Abbas Mohamed for supervising my thesis. I deeply appreciate his support, his guidance, his helpfulness throughout the thesis process and the several and precious suggestions he has given me for my future professional choices.

In this research, I was lucky to have Dr Paolo Paron in the supervision team. I would like to express my appreciation to him for giving me the opportunity to gain a lot of knowledge from his personal experience and for guidance, helpfulness and suggestions throughout the thesis process. I would like to thank him for joining me the field campaign in Roseires Reservoir in a very difficult weather using truck.

I also extend deepest gratitude for the Msc students Amgad Omer and Sivia Zini who conducted their Msc thesis under my PhD research. Their findings were contributed to the success of this thesis.

I extend my appreciation to the Ministry of Water Resources in Ethiopia for providing data free of charge and allowing me conducting a famous fieldwork along Blue Nile River network. Their support was one of the main reasons of the success of this research, thanks a lot Samunish and Biruk Kebede from the hydrology department and Abiti Getaneh, the director research, Ministry of Water Resources Ethiopia.

I would like to thank Eastern Nile Technical and Regional Office (ENTRO) in Ethiopia for offering me rubber boat with engine to assist me during the measurement and providing me the available data.

I appreciate my colleagues at Hydraulic Research Center of the Ministry of Water Resources and Electricity for their support during all stages of this research including data collection and giving comment and ideas for my research work. Special thanks to Mr Mohanad Osman for joining us in the difficult fieldwork in Roseires Reservoir in Sudan, Mr Abu Obeida Babiker Ahmed for his valuable advises during the study and Mr Adil Dawoud for analyzing more that 500 sediment and water samples that I have collected from Ethiopia and Sudan.

I also extend deepest gratitude for the staff of Dam Implementation Unit (Roseires Dam Heightening) for supporting my extensive fieldwork, willingness to give me the data I required and kindness during my fieldwork in Roseires Reservoir. Special thanks to Mr. Yasir Abo El Gasim, Eng. Hussien and Eng. Khidr for their continuous support in the fieldwork and social interaction.

I would like to express my appreciation to Dr. Kees Sloff from Deltares for continues guidance in the morphodynamic model (Sobek River/mMorphology) and providing important assistance and opinions on this thesis.

I am very grateful to Mahmood Rabani Foundation for financial support of the field work in Roseires Reservoir and the laboratory analysis.

I would like to express my deepest thankful and respect for Jolanda Boots, the PhD administration officer who did all the possible and impossible to make success of my study.

I have been very fortunate to have many friends who always supported me and shared nice times together with me. Thank you my friends: Chol Abel, Khalid El Nour, Girma, Fekadu, Sirak, Hermen,Rahel, Eshraga, Abonesh, Dr Melesse, Reem, Eman, Shaza, Marmar, Salman, Sami and Mohanad. Ermias Teferi, I give my special thanks to you, for always helping me in so many things and for creating a very comfortable and convenient and for being my good and wise friends. Thanks also to all other friends who I haven't mentioned but who have helped me and encouraged me.

I would also like to express my gratitude to all committee members for reading my thesis and giving me great comments for improving my thesis.

I would like to express my deepest thankful and respect for Sudanese friends at home and Delft for their priceless support, continuous advice and communication which was a very important and inspiring for my study.

Finally, to my family, 1 would like to express my gratefulness for their endless source of love, inspiration, patience and understanding during all these years. Along with our beloved daughters, Yara and Yem, both make every day a living dream

TABLE OF CONTENTS

Chapter 1
INTRODUCTION

1.1 BACKGROUND

The Blue Nile River Basin is increasingly under human pressure, due to rapidly growing population in Sudan and Ethiopia (Balthazar et al., 2013). This has already resulted in a number of environmental problems caused by the extensive exploitation of territory and resources (Garzanti et al., 2006).

Rapid population increase in the upper Blue Nile led to fast land-use changes from natural forest to agricultural land, which resulted in speeding up the soil erosion process (Figure 1.1 (a)). Slope failures of the deep gorges and rugged valley walls which caused land sliding and rock falling is another factor leading to soil erosion in the basin (Ayalew and Yamagishi, 2004). Soil erosion undesirably increases the sediment load downstream and reduces soil fertility and hence agricultural productivity upstream. Eroded sediment particles are transported away by the flowing water with undesirable downstream sedimentation as a result.

In the reservoirs, sedimentation leads to serious reduction in storage capacity, causing hydropower generation problems and negative impacts on society and economy, environment and ecology (Abdalla, 2006). The sediment settling inside the irrigation canals leads to water shortage and management difficulties (Figure 1.1 (b)). Sediment deposition on the bed of the river raises the bed level and enhances flood risks. However, sedimentation has also some positive impacts: on the river flood plains, the settled fine sediment acts as fertilizer for agriculture. Research showed that each tonne of sediment passed to the fields is equivalent to 0.94 kg of urea fertilizer (ENTRO, 2007; Gismalla, 2009). Another benefit of sediment deposition is brick making manufacture practiced along the river and the irrigation canals banks (Gismalla, 2009). Nevertheless, the negative impacts of sedimentation are larger than the positive ones, so we can consider sedimentation as a problem rather than a benefit for the Blue Nile River basin (Gismalla, 2009).

| (a) | (b) |

Figure 1.1. Soil erosion in the upper basin in Ethiopia (a) and hills of sediment deposition along irrigation canals in Sudan (b).

Mitigations of soil loss problem necessitate erosion control practices in the upper catchment. This problem has a trans-boundary character, since it needs the cooperation of the two riverine countries Ethiopia and Sudan. Given the vastness of the region, it is important to identify the source areas delivering the highest quantities of sediment to the downstream sinks. The analysis of land-use changes and the identification of the degree of soil erosion are not enough to identify these areas. So, a study that combines source to sink through sediment transport in the river system is needed.

The Blue Nile River Basin is currently experiencing new developments, both in Ethiopia and in Sudan. The Grand Ethiopia Renaissance Dam (GERD) is under construction about 30 km upstream of the Ethiopian-Sudanese border (Ali et al., 2013a). Recently, the Roseires Dam, located 110 km downstream the Ethiopian-Sudanese border, has been heightened by 10 m, increasing the storage capacity of the reservoir by additional 3700 million m^3 (Ali et al., 2013b). Other mega dams are planned in Ethiopia for hydropower production. These developments will strongly affect the water resource and sediment deposition in the lower Blue Nile Basin (Abdelsalam and Ismail, 2008). Therefore, it is needed to take into account these developments in a study that combines source to sink of sediment for the future situation.

1.2   RESEARCH AIMS

The goals of the research are to identify the most critical eroding areas in the upper basin and to study the effects of the new and planned developments on the sedimentations rates downstream. This was done by studying characteristics and amounts of the material deposited in the first sediment trap of the river, Roseires Reservoir. Tracing back the sediment source area was conducted by comparing the mineral characteristics of deposited sediment in the reservoir with the soil in the upper basin. Models allow studying sediment transport and water distribution and assess the effects of the new developments are the major tools used in this research. This research therefore combines the analysis of historical data, numerical model results, laboratory analysis of sediment samples and conceptual modelling. The study includes also intensive field measurement campaigns to collect the necessary data. This study provides an answer for the following questions:

> **Does the erosion in the upper Blue Nile River catchment in Ethiopia result in increased sediment load and increased sedimentation in Sudan?**

It is believed that most of the sediment material settling in Sudan originates from soil erosion in the upper catchment area in Ethiopia. This needed to be studied further particularly to determine from which sub basins the sediment comes. This high sediment load influences the design and operation of the reservoirs and the irrigated schemes. It was therefore important to establish the contribution of erosion in the upper catchment to plan the implementation of mitigation measures that are effective and targeted.

> **What are the hydromorphological and sedimentological characteristics of the Blue Nile River system in Ethiopia and Sudan?**

The Blue Nile River brings considerable amounts of sediment during the flood season. The transported sediment in the river consists of significant quantities of fine material (silt and clay) which can be easily transported in suspension (Hussein and Yousif, 1994). Suspended sediment accounts for approximately 90% of the total sediment load in the river.

The hydrodynamic characteristics of the river network in Ethiopia and Sudan were not known and also the amounts and types of the sediment transported by the river network. Therefore, it is important to study the sediment transfer along the river network and to identify the source of the sediment deposited in reservoirs.

> **What will be the impact of the Grand Ethiopian Renaissance Dam (GERD) and heightening of Roseires Dam on sediment processes?**

Sediment deposition inside Roseires Reservoir has reduced the reservoir storage capacity by one third already, which has resulted in shortage in irrigation water for agriculture schemes and reduction of hydropower generation. Although the dam heightening has increased the storage capacity of the reservoir, but it will result also in increasing sedimentation rates.

The construction of Grand Ethiopian Renaissance Dam (GERD) in Ethiopia will have significant impacts on the downstream sediment transport. This study has clarified the role of heightening and GERD on sedimentation rates inside Roseires Reservoir.

1.3   THE BLUE NILE HYDROSOLIDARITY PROJECT

This research is a part of a larger project aiming to enhance the understanding of the hydrological and biophysical processes in the Blue Nile River Basin in order to quantify the relationships between land-use management upstream and water availability and sediment loads downstream. The project name is: "*In Search of Sustainable Catchments and Basin-wide Solidarities; Trans-boundary Water Management of the Blue Nile River Basin*". The project is funded by WOTRO; Science for Global Development programmes of the Netherlands Organization for Scientific Research (NWO). There are 11 research components addressed in this project as illustrated in Figure 1.2.

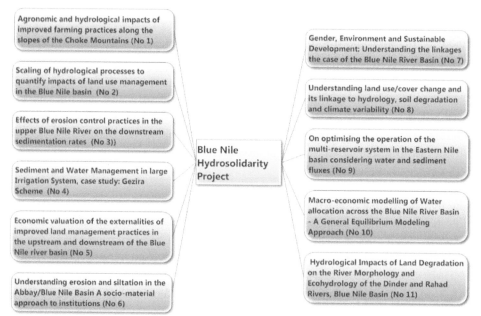

Figure 1.2. Reseach topics addressed in Blue Nile Hydrosolidarity Project.

The main objective of the project is to enhance the collaboration between Dutch, Ethiopian and Sudanese knowledge institutes concerning hydrology and river basin management, as well as to strengthen the mutual understanding and solidarity between the countries riparian to the Blue Nile River Basin (van der Zaag and Belay, 2007).

This research in this thesis (No, 3 in the overall project) has direct or indirect interaction with other researches within the framework of the overall project, as shown in Figure 1.3.

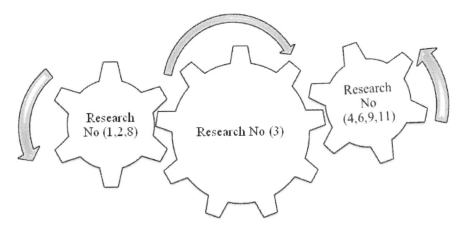

Figure 1.3. Interaction between this research and other research within the overall project.

1.4   GENERAL METHODOLOGY

Roseires Reservoir has already lost about one third of its storage capacity in about 50 years. Sedimentation inside the reservoir could be mitigated by appropriate management of the upper basin. To do that, the areas providing the highest sediment volumes to the river have to be identified, since they should have priority with respect to the application of erosion control practices. The identification of the source of the sediment depositing in the lower Blue Nile River is based on the study of the sediment transfer from the source in the upper basin to the sinks in the lower basin and relative sediment budgets. The methods used in this research aim at answering the following questions:

> **How can we determine sediment origin and timing of sediment transfer along the Blue Nile River?**

The origin of the sediment depositing in Roseires Reservoir can be identified based on the comparison of the mineralogical characteristics of the sediment layers deposited inside the reservoir and the eroding material in the upper basin. The identification of the time of deposition can be derived from soil stratification which leads to the identification of specific years. The most promising coring locations inside the reservoir can be identified from a combination of data analysis and model results as areas that are not subjected to erosion or formation of alternate bars. The mineral characteristics of sediment samples collected from the selected locations can be analyzed using X-ray powder diffraction to determine the mineral content. Cluster analysis then allows performed grouping the samples with similar mineral contents.

The study of the sedimentation rates and identification of origin requires the knowledge of the water and sediment balances along the river system.

> **How can we estimate the water and sediment balances along the Blue Nile River Basin both in Ethiopia and Sudan?**

The promising method to assess the flow balance is integrating the measured flow at several gauging stations with the results of the Soil and Water Assessment Tool (SWAT), a basin-scale model allowing estimating the water and sediment in un-gauged basins.

Assessing the sediment budgets requires analysis of flow and sediment data. The suspended sediment rating curve presenting the relationship between flow discharge and either suspended sediment concentration or suspended sediment load could be used to predict unmeasured sediment concentrations and sediment loads from measured discharge at the time.

Long-term land-use and land-cover changes (LULCC) detection could be performed through the analysis of Land sat images from different years to confirm the sediment product with land-use land cover changes. Several pre-processing methods could be implemented to prepare the land-use maps for classification and change detection including geometric correction, radiometric correction, topographic normalization and temporal normalization.

The flow and sediment data are the key issue to develop a model capable of following the sediment from their entrance in the river system to their final sinks.

> **What types of methods and tools can be used to follow the sediment to the sinks?**

Many tools can be used to simulate the large scale sediment transport in rivers but these models are mainly designed for non-cohesive sediment transport with the capacities to simulate simple processes of cohesive sediment transport. These tools include HEC-6 (Army, 1993), GSTARS 2.1, GSTARS3 (Yang and Simaes, 2000; Yang and Simaes, 2002), GSTAR-I D (Yang et al., 2004-2005) and Sobek-RE (Sloff, 2007) among others.

In river systems, where more than 90% of the sediment transport is fine material, a modelling system including advection diffusion equation for suspended sediment appears promising to simulate the fine sediment transported by the Blue Nile River.

> **How can we develop a basin scale model for Blue Nile capable of sediment transfer?**

The derived flow and sediment balances combined with intensive field measuring campaigns including soil bed material, river cross sectional profile and reservoirs operation policy allows constructing a model to simulate the morphodynamic processes of the Blue Nile River system. The most promising tool is the Sobek Rural-Delft3D-Delwaq (Sobek River/Morphology) software capable of simulating the processes of hydrodynamics and sediment transport. Sobek morphology allows simulating the silt and sand transport along the Blue Nile River from Lake Tana (source) to Roseires Dam. The model will be calibrated and validated based on the Roseires Reservoir bathymetric surveys and sediment measured at the Ethiopian-Sudanese border.

> **How can we assess the new developments along Blue Nile River System?**

The calibrated model can be used to assess the effects of the new developments along the river system such as Roseires Dam heightening and Grand Ethiopian Renaissance Dam after construction. The model can be finally used also to assess the impact of different erosion control practices on the sedimentation rates inside these reservoirs.

1.5 STRUCTURE OF THE THESIS

This research is outlined as follows:

Chapter 1 introduces the general problem in the Blue Nile Basin, and provides a brief description of the Blue Nile Hydrosolidarity Project and its components. Finally, the chapter presents the main objectives of the thesis and the general methodology used to achieve the objectives.

Chapter 2 describes the river system in detail, including location, topography, geology, climate, hydrology, sediment transport and the hydraulics structures constructed along the river system. The chapter focuses on the reservoirs sedimentation of Roseires and Sennar dams.

Chapter 3 quantifies the river flows and sediment loads along the river network. The Soil and Water Assessment Tool was used to estimate the water flows from un-gauged sub basins and to compare the estimated sediment loads at selected locations. For the gauged sub basins, water flows and sediment loads were derived based on the available flow and sediment data using rating curves.

Chapter 4 presents the results of a one-dimensional hydrodynamic model covering the entire Blue Nile River system with the aim to quantify the water availability throughout the year for different conditions.

Chapter 5 presents the analysis of sedimentation processes inside Roseires Reservoir. Delft3D modelling and field surveys of both sediment quality and quantity were used to identify the areas that were neither subject to net erosion nor to bar migration during the life span of the reservoir, and these locations were selected for sampling. The results allowed linking the sediment deposited inside Roseires Reservoir (sink) with the eroded soils and rocks in the upper basin (source) using x-ray diffraction to identify the mineral content and cluster analysis modelling.

Chapter 6 shows the sediment transport along the entire Blue Nile River using the modelling system SOBEK-River/Morphology to perform morphological simulations for river system. The model was used further to assess the morphological impacts of the planned developments in the basin.

Chapter 7 draws the discussion and conclusions.

Chapter 2
BLUE NILE RIVER BASIN

Summary
This chapter will describe the Nile Basin River network including its sub basins. It will then focus on the Blue Nile River basin; the case study of this research. The history of sediment deposition inside Roseires and Sennar reservoir will be coved in this chapter.

2.1 NILE RIVER

The Nile is the longest river in the world, with a total length of about 6,700 km and a relatively small basin area, about 3×10^6 km², although accounts range between 1,878 and $3,826\times10^6$ km² (Milliman and Farnsworth, 2011). The Nile valley is the cradle of important ancient civilizations, thanks to the regular discharge regime of the river, characterized by a single flooding season, and the fertilising effects of the silts deposited on floodplains. For this, the Nile has a long history of human interference, evolving from water withdrawal for irrigation, to canalization (mainly in the delta region) and damming (Williams, 2009). Today, the Nile is an important trans-boundary river flowing through eleven countries: Burundi, Rwanda, Tanzania, Kenya, Uganda, Democratic Republic of Congo, South Sudan, the Sudan, Ethiopia, Eritrea and Egypt (Figure 2.1). For this, and considering also that the Nile is the main fresh water resource for the Sudan and Egypt, the management of its waters has an international character with potential sources of conflict (Allan, 2009).

The Nile River has three major tributaries, the White Nile and the Blue Nile, merging into the main Nile at Khartoum, and the Atbara River, joining the river at Atbara, in the Sudan. Taking this into account and for sake of convenience, the description of the river network is here subdivided in five parts, from south to north namely; White Nile River, Blue Nile River, Atbara River, Main Nile River and Nile Delta.

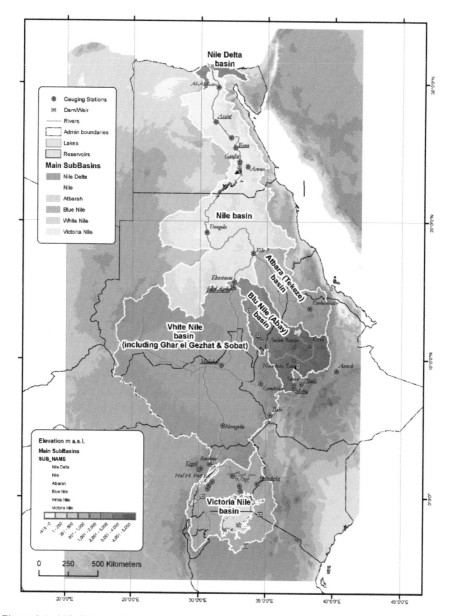

Figure 2.1: Nile River basin and main sub basins.

2.1.1   White Nile River

The river is the longest tributary of the Nile, approximately 3,700 km long, but it contributes to only 20 to 30% of its annual flow, with an average annual discharge of about 28 billion m³/year. The river headwaters flow through four major lakes in the lake plateau of Eastern Africa, including Lake Victoria. The river flows down to the vast seasonal swamps of the Sudd, in South Sudan, one of the largest freshwater wetlands in the world (Green and El-Moghraby, 2009). Leaving the Sudd, the river meets the Bahr el Gazal, flowing from west, at Lake No, forming the White Nile River (Sutcliffe, 2009), and then the Sobat River, flowing from the Ethiopian highlands. The White Nile merges with the Blue Nile at Khartoum, the capital city of Sudan, forming the Main Nile.

2.1.2   Blue Nile River

The Blue Nile is the tributary providing most of the waters to the Main Nile, 60-65 % of its total annual flow, with a clear bimodal seasonal pattern and 72-90 % of its sediment yield (Goudie, 2005; Williams and Talbot, 2009). The river originates in Ethiopia, on the western side of the Main Rift Valley flanks, from Lake Tana (Shahin, 1985). The river flows for nearly 1635 km to Khartoum, where the river meets the White Nile to form the main Nile River. The river and its sub basins are described in detail in the next sections.

2.1.3   The Atbara River

The river is the major tributary of the Main Nile River, joining it 320 km north of Khartoum at the city of Atbara (Shahin, 1985). With an average annual discharge of almost 12 billion m^3, the river accounts for 10 to 15 % of the total Nile's flow and for 25 % of the sediment yield. The Atbara is 800 km long and has its source in the northwest Ethiopia highlands. The Atbara has three major tributaries, all with a strong seasonal character. The Tekezé River, the largest one, 600 km long, is dry for nine months per year. A huge hydroelectric dam (185 m high), called the Tekezé Dam, completed in 2009 in Ethiopia, where the river flows through a narrow and steep canyon, up to 2,000 m deep. The Atbara is crossed by the Khashm El Girba Dam, near Kassala in Sudan, constructed in 1964 to store water for the Halfa irrigation scheme. The Ministry of Water Resources and Electricity of Sudan is constructing Dam complex of Upper Atbara Projects. The project encompasses of Rumela Dam, situated on Upper Atbara River and Burdana Dam, situated on Setit.

2.1.4   The Main Nile River

The river starts at Khartoum, and flows for about 3,000 km through the Sahara desert forming a green fertile corridor to the Mediterranean Sea, in Egypt. In its first part, the Nile River forms a large meander, extending from Atbara to Dongola, which is determined by geological constraints, due to the presence of ancient faults and plutonic domes (Talbot and Williams, 2009). Before entering Egypt, the Nile flows to Lake Nasser (called Nubia in its 200 km long Sudanese part). From the lake outlet, near Aswan, to Cairo, the capital city of Egypt, the river flows for 1,600 km. North of Cairo, the Nile splits in the two branches of Rosetta and Damietta forming a wide delta.

2.1.5   The Nile Delta

The delta was formed about 8,000 years ago, after the rapid rise of sea level caused by global warming after the last glacial maximum. It extends for 22,000 km², its apex being located 23 km north of Cairo (200 km from the shoreline at 18 m (AD)), where the Main Nile bifurcates in the two branches of Rosetta and Damietta. Two thousand years ago there were at least five distributaries, forming small promontories, which continued to transport significant volumes of sediment during the annual floods. Since then, humans significantly influenced the delta evolution, especially by constructing a dense network of irrigation channels, which currently trap virtually all fine sediment transported by the river (Stanley, 1996).

2.2   BLUE NILE RIVER AND ITS BASIN

2.2.1   Location and topography

The Blue Nile River has total drainage area of approximately 330,000 km² (Peggy and Curtis, 1994), and contributes approximately 62% to the flows in the Nile River (Waterbury, 1979; Yates and Strzepek, 1998 b). The basin is characterized by considerable variation in altitude, ranging from 367 m (AD) at Khartoum to 4,256 m (AD) above sea level in the Ethiopian highlands (Figure 2.2 a).

The source of the Blue Nile is a small spring at an elevation of 2,900 m above sea level, about 100 km to the south of Lake Tana. From this spring, the Little Abay River (Gilgile Abay) flows down to Lake Tana. The Lake is the biggest lake in Ethiopia; about 73 km long and 68 km wide. It is located at 1,786 m (AD) and has a surface area of 3,042 km² and stores 29.2 billion m³ of water which fluctuates seasonally between 1,785 and 1,787 m (AD). The lake is shallow and has a mean depth of 9.53 m, while the deepest part is 14 m. From Lake Tana, the river travels 35km to the Tissisat Falls, where it drops by 50 m (Awulachew et al., 2008). The river then flows for about 900 km through a gorge crossing the Ethiopian Highlands, which in some places is 1,200 m deep.

The basin is divided into 17 major sub basins namely, Blue Nile Sudan, Dinder, Rahad, Tana, Jemma, Beles, Dabus, Didessa, Jemma, Muger, Guder, Fincha, Anger, Wonbera, South Gojam, North Gojam and Welaka (Figure 2.2 b). The Dinder and Rahad rise to the west of Lake Tana and flow westwards across the border joining the Blue Nile below Sennar Dam. The major rivers in each sub basin are summarized in Table 2.1.

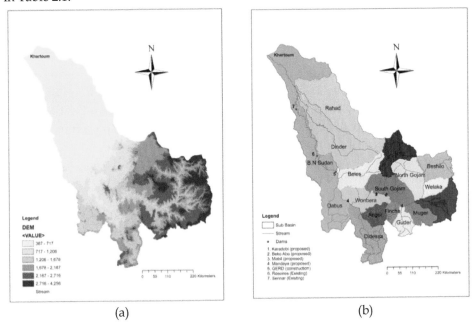

(a)   (b)

Figure 2.2: Blue Nile River- topography (a) and sub basins and dams (b).

Table 2.1: Blue Nile Sub basins characteristics.

No	Sub basin name	Area (km²)	Major rivers
1	Tana	15142	Ribb, Gumera, Megech
2	Jemma	13362	Jemma
3	Beles	14146	Main Beles, Gilgel Beles
4	North Gojam	14389	Andessa, Mendel, Muga, Shina
5	Dabus	21252	Dabus, Hoha, Haffa
6	South Gojam	16762	Jedeb, Birr, Chamoga, Temcha, Fettam, Dura
7	Jemma	15782	Gumero, Jemma, Winchit
8	Welaka	6415	Jogola, Mechela, Selgi
9	Wonbera	12957	Belzmate
10	Fincha	4089	Neshi, Fincha
11	Anger	7901	Little Anger, Anger
12	Muger	8188	Aletu, Muger
13	Didessa	20087	Dabana, Didessa
14	Guder	7011	Debis, Guder
15	Dinder	37611	Dinder
16	Rahad	42303	Rahad
17	B.N Sudan	52999	Blue Nile

2.2.2 Climate

The climate in the basin varies considerably between the Ethiopian highlands and its confluence with the White Nile in Khartoum. The spatial and temporal variation is affected by the movement of air masses associated with Inter Tropical Convergence Zone (ITCZ). During the winter dry season (known in Ethiopia as Bega) the ITCZ lies south of Ethiopia and the Blue Nile region is affected by a dry northeast continental air-mass. From March, the ITCZ returns bringing Small rains (known in Ethiopia as the Belg) particularly to the southern and south western parts of the Basin. In May, the northward movement of the ITCZ produces a short intermission before the main wet season (known in Ethiopia as the Kremt). Around June, the ITCZ moves further north and the southwest airstream extends over the entire Ethiopian highlands to produce the main rainy season. This is also the main rainy season in Sudan, though being further north and at lower altitude (Awulachew et al., 2008).

The average annual rainfall varies between 1400 and 1800 mm/year, ranging from an average of about 1000 mm/year near the Ethiopia–Sudan border to 1400 mm/year in the upper part of the basin, and in excess of 2000 mm/year in the parts Didessa and Beles sub basins (Awulachew et al., 2008). In Sudan, the rainfall drastically decreases from about 1000 mm/year near the border with Ethiopia to less than 200 mm/year at the junction with the White Nile in the city of Khartoum (Gamachu, 1977; Sutcliffe and Parks, 1999). The rainfall trend was studied in the upper Blue Nile basin, most of the results showed that, there was no significant trend in the seasonal and annual basin-wide average rainfall (Conway, 2000; Seleshi and Zanke, 2004; Tesemma et al., 2010). The mean monthly rainfall for representative climate stations located in Blue Nile Basin both in Sudan and Ethiopia is shown in Table 2.2.

The annual mean potential evapotranspiration decreases with increasing elevation form 1845 mm to 924 mm (Conway, 1997). In the Sudan, potential evaporation increases, this produces a significant loss of Blue Nile water. For instance, the Sennar region has a potential evaporation rate of 2,500 mm/year, but receives only 500 mm/year of rain (Block, 2007; Shahin, 1985).

The temperature in the basin varies with elevation. The climate is generally temperate at higher elevations and tropical at lower elevations. Traditional classifications of climate in the upper basin use elevation as a controlling factor and recognize three regions namely; the Kolla zone below 1800 m with mean annual temperatures in the range 20-28°C, the Woina Dega zone between 1800-2400 m with mean annual temperatures in the range 16-20°C and the Dega zone above 2400 m with mean annual temperatures in the range 6-16°C (Conway, 1997). Increasing trends in temperature has been reported at different weather stations in the upper Blue Nile in Ethiopia (Tekleab et al., 2013). The analysis in the period 1941–1996, shows an important piece of evidence of warming in most parts of the Sudan, namely central and southern regions (Elagib and Mansell, 2000).

Table 2.2: Mean monthly rainfall (mm) for representative climate stations located in Blue Nile Basin.

Station	Jan	Feb	Mar	Apr	May	June	Jul	Aug	Sep	Oct	Nov	Dec	Annual
						Ethiopia							
Gonder	5	4	19	34	87	151	311	279	116	56	24	9	1,095
Bahridar	3	2	8	22	83	181	444	395	196	92	23	4	1,453
Debre Marcos	12	22	49	61	96	155	301	300	204	81	24	15	1,320
Debre Tabor	6	11	42	46	93	180	501	476	193	66	21	16	1,651
Addis Ababa	19	50	71	90	95	118	250	265	169	41	9	14	1,191
Sibu Sire	16	24	50	85	150	220	271	245	179	74	42	14	1,370
Jemma	30	52	95	128	166	209	213	210	183	93	66	32	1,477
Gore	39	46	98	119	234	327	336	335	337	166	102	37	2,176
Gambela	5	8	27	56	156	150	239	228	155	113	48	12	1,197
Assossa	0	0	31	32	118	189	207	208	207	103	21	0	1,116
						Sudan							
Hawata	0	0	0	2	12	97	142	210	82	20	1	0	566
Rosieres	0	0	1	12	36	119	157	175	121	33	1	0	655
Sennar	0	0	0	3	14	60	113	143	75	25	1	0	434
Wad Medani	0	0	0	1	13	28	88	111	46	16	2	0	305
Khartoum	0	0	0	0	4	7	43	63	17	5	1	0	140
Rahad	0	0	0	2	25	37	138	130	71	24	1	0	428

Source: FAO-CLIM2 Worldwide agroclimatic database

2.2.3   Hydrology

The flow of the river reflects the seasonality of rainfall over the Ethiopian highlands, where there are two separate periods. The flood period, or wet season, extends from July to October, with maximum flow in August-September and the low flow, or dry season, takes place between November and June

The average annual flows of the river downstream Lake Tana (1959–2003) and Kessie Bridge 1953–2003) was estimated to be 3.9 billion m^3/year and 16.3 billion m^3/year respectively. The long-term (1912–2010) mean annual discharge of Blue Nile entering Sudan and measured at El Diem (Figure 2.3) is 49.29 billion m^3/year with maximum in 1929 and minimum in 1984 (Abdelsalam and Ismail, 2008). River flow analysis using statistical test at El Diem station at Ethiopian–Sudanese border showed significant increasing trend during the main rainy season (Gebremicael et al., 2013; Tesemma et al., 2010).

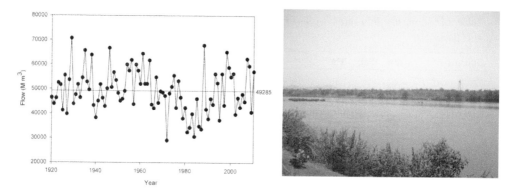

Figure 2.3: Hydrograph of the Blue Nile at El Deim 1920-2010 in Mm³ (left) and El Deim Station (right).

Since 2001, the outflow from Lake Tana has been regulated by construction of the Chara Chara Weir for hydropower generation. This has resulted in a change in the natural pattern of flow from the lake, with higher dry season flows and lower wet season flows (Figure 2.4). However, because the flow from Lake Tana is a relatively small proportion of the flow at El Deim (7%), the regulation is not thought to have had a significant impact on the distribution of flows downstream

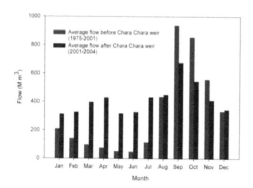

Figure 2.4: Modification of outflow from Lake Tana as a consequence of regulation by the Chara Chara Weir (left) and Chara Chara Weir (right).

Dinder and Rahad tributaries join the Blue Nile River Downstream Sennar Dam and their annual average (1980-2010) contribution is about 2.0 billion m³ /year and 1.1 billion m³ /year respectively

2.3   STRUCTURES AND WATER EXTRACTION FOR IRRIGATION

The Blue Nile system encounters several structures from its source to Khartoum. The existing system consists of small structure such as Chara Chara Weir and Fincha Dam and big structures such as Roseires and Sennar reservoirs.

2.3.1   Chara Chara Weir and Fincha Dam

Chara Chara Weir was built in 1998 at the outlet of Lake Tana to regulate the flow for hydropower production at Tis Abay I and Tis Abay II power stations (Awulachew et al., 2010; McCartney et al., 2009).

Fincha Dam was built across the Fincha River, a tributary of the Blue Nile, in 1972 to regulate the river flow for hydropower production and sugarcane irrigation (8145 ha) with a reservoir capacity of 400 million m³(Awulachew et al., 2008).

2.3.2   Roseires Reservoir

Roseires Dam (Figure 2.2 b) is built for irrigation and hydropower generation (completed in 1966). The dam consists of five deep sluice gates, each measure 6 m wide by 10.5 m high at an invert level of 438.5 m (Alexandria datum (AD)). Spillway at a crest level of 466.7 m (AD) was constructed using 10 radial gates, each measure 12 m width by 10 m high (Hussein et al., 2005). The hydropower station is located on the flood plain. The originally designed reservoir at 484 m (Alexandria datum) has a length of 75 km giving a total capacity of 3024 million m³ at the maximum impoundment level of 484 m. This reservoir is the first sediment trap for the sediments transported by the river. Due to sedimentation, it has lost a storage volume of approximately 1,000 million m³ already (Ali and Crosato, 2013).

The dam was recently heightened by 10 m in a second stage which resulted in increasing the old storage capacity of the reservoir by approximately additional 3.7 billion m³. Figure 2.5 illustrates the reservoir boundaries before and after the dam heightening with some photos from downstream the dam. The new maximum operation level is now 493 m above sea level according to Alexandria datum (AD).

Before the heightening, Roseires Reservoir was used to be maintained at a level of 470 m (AD) which was the lowest operating level during the rising flood. Over this operation period, minimum sediment deposition is expected despite the large quantities of sediment inflow, which may reach more than 3.0 million tonne/day (Bashar and Eltayeb, 2010). The reservoir filling period commences after the flood peak has passed. According to Roseires Reservoir Operation Rules before heightening, filling may start any time between the 1st and the 26th of September each year depending on the magnitude of the flow at El Deim. Atypical operation program before heightening for a median year is shown in Figure 2.6 (Ahmed et al., 2010; Bashar et al., 2010).

Figure 2.5: Roseires Reservoir surface area before and after dam heightening (left), photo for the dam before heightening (right-upper) and during heightening (right-lower).

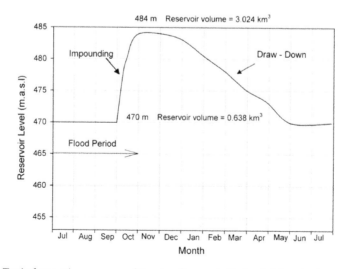

Figure 2.6: Typical operation program of Roseires Reservoir (before heightening).

In 2010, a topographic and bathymetric survey was carried out at the new operation level of the reservoir after heightening; the results of variation in storage volume and surface area with elevation are shown in Figure 2.7.

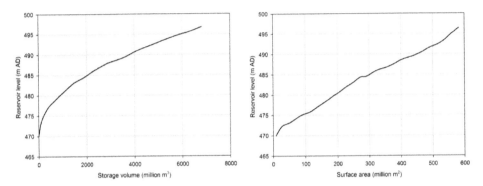

Figure 2.7: Roseires Reservoir variation of storage volume (left) and surface area (right) with elevation after dam heightening.

The active storage volume between the new minimum operation level (470 m AD) and the new maximum operation level (493 m AD) is 5985 million m³ according to the bathymetric survey of 2010 (SMEC, 2012). The new operation program is discussed in chapter 6.

2.3.3   Sennar Reservoir

Sennar Dam (Figure 2.2 b) is the first reservoir constructed along the Blue Nile in Sudan (1925) some 350 km south Khartoum, with a storage capacity of 0.93 billion m³ (Awulachew et al., 2008). The main purpose of the dam is to irrigate the Gezira Scheme and secure drinking water supply during the dry season. The main section of the dam (Figure 2.8) is a masonry wall of 1600 m long and 30 m maximum height. Including the earth fill embankment sections on both banks; the total length of the dam becomes 3 km. The dam contains eighty low level sluices which are adequate to pass the normal seasonal floods; the dam is also provided with spillways at higher level to pass the peaks of exceptional floods. Head regulators for the Gezira and Managil canals are situated at the west end of the masonry section. During the peaks of the floods the reservoir is held at 420.2 m (AD), a level corresponding to the sills of the spillway, and is subsequently filled on the falling flood, when the sediment content of the river inflows has reduced. The total capacity of the reservoir at maximum reservoir level of 424.7 m (AD) has been reduced to 328 million m³.

The operation of Sennar Dam is similar to Roseires Dam. A sluice gates and spillway are fully opened during the high flood to maintain the water level in the reservoir at minimum of 420.2 m (AD) to reduce siltation. The reservoir filling is carried out based on the falling flood according to the flow at El Deim upstream of Roseires

Reservoir and then follows a day program. The starting for filling lies between 1st September and 26th September to a maximum level in the reservoir.

Figure 2.8: Sennar Dam durning flood season, upstream (left) and downstream (right).

2.3.4 Proposed Dams

A major dam named Grand Renaissance (Millennium) Dam (Figure 2.2 b) is currently under construction in Ethiopia, 30 km upstream of the Ethiopian-Sudanese border. It will be the largest hydroelectric power plant in Africa with a reservoir capacity of 63 billion cubic meters (the dam speech, 2011). The United States Bureau of Reclamation carried out a major study of the land and water resources of the Blue Nile River basin in Ethiopia over the period 1960-1964. The study identified major hydropower development sites on the main stream of the Blue Nile. These dams are Border, Karadoby, Mabil and Mandaya (Figure 2.2 b). The Eastern Nile technical and regional office (ENTRO) has carried out pre-feasibility studies for Karadobi, Mandaya and Border in 2006 and 2007. In their study, they changed the storage capacity of Mandaya to 49.2 billion m^3 and Boarder dam to 13.3 billion m^3 instead of 11.1 billion m^3, while the Mabil dam was cancelled. Moreover, they have proposed a potential of hydropower generation in Beko Abo. The general characteristics of these projects are presented in Table (2.3).

Table 2.3: The characteristics of proposed hyropower projects on the Blue Nile River (USBR and ENTRO).

Dam name	Border	Mandaya	Mabil	Karadobi	Beko Abo
Dam height (m)	84.5	164	171	250	282
Full supply level (m.a.s.l)	575	741	906	1146	1062
Capacity Million m3)	11074	15930	13600	40200	31700
Design power (MW)	1400	1620	1200	1600	1940

2.3.5   Water extraction

Several agricultural schemes are irrigated using the waters of the Blue Nile River (Figure 2.9). The largest one is the Gezira-Managil Scheme (operating since 1925), totalling about 880,000 ha, followed by the Rahad (since 1979) and the Suke (since 1971) schemes, covering 126,000 and 37,000 ha, respectively. The North West Sennar Sugar Scheme (since 1974) and the Genaid Sugar Scheme (since 1961) cover 20,000 and 15,960 ha, respectively. Additionally, many small irrigation schemes supplied by pumping directly from the river are found along the river banks. The extraction period and the amount of water are clearly indicated in the water rights issued. Their extraction volumes however, are unknown.

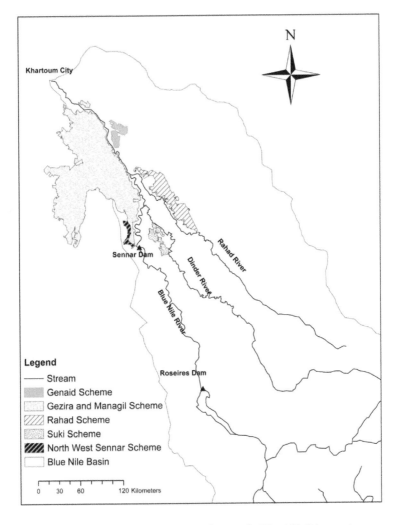

Figure 2.9: Locations of agriculture schemes irrigated using the Blue Nile River waters.

2.4 SEDIMENT TRANSPORT

The transported sediment in the Blue Nile consists of significant quantities of very fine material composed of silt and clay with diameter less than 63 microns known as wash load which can be easily transported in suspension, and under certain hydraulic conditions it is ready to settle fast (Hussein and Yousif, 1994).

Suspended sediment transport at El Deim Station near the Ethiopian Sudanese border is reported to be 123 million tonne/year (El Monshid et al., 1997; Siyam et al., 2005). Bed load has been estimated in 15 % of the suspended sediment loads, giving a total mean annual sediment transport of 140 million tonne/year (El Monshid et al., 1997; Siyam et al., 2005).

10 days average measured discharge and sediment concentration at El Deim and the total rainfall over the basin (Figure 2.10) showed that the peak of the discharge (513 Million m³/day), comes after the peak of the sediment (5660 ppm) by about three weeks, while the peak of the total rainfall (2123 million m³/day) comes one week before the peak of the discharge. This sequence of incidents is logical. It is attributed to the fact that at the beginning of the rainy season, the catchment area is bare, there is little green vegetation and the soils are exposed and easily eroded. The time lag between these peaks depends on the rainfall intensity, duration and temporal distribution as well as on the condition of the catchment.

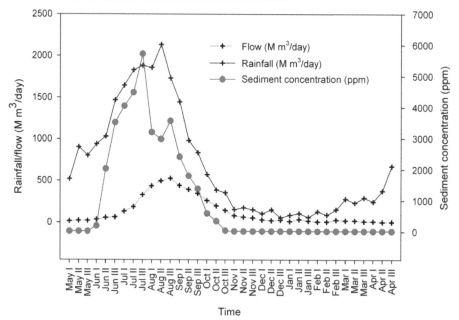

Figure 2.10: Comparison between 10 days average rainfall, discharge and sediment yield at El Deim.

Figure 2.11 (a, b and c) shows the sediment concentrations measured in the last three decades at El Deim. Averaged values of the concentrations were derived for three decades: 1970s, 1990s and 2000s to represent incoming sediment concentrations during high flow (Omer, 2011). Previous studies indicated differences in concentration between the rising limb and the falling limb of the flood curve (Ahmed and Ismail, 2008; Billi and el Badri Ali, 2010) , but with high variability of concentrations.

The correlation between river flows and the sediment discharges for the rising and falling flood limbs at El Deim station for the period 1970 – 2010 is depicted in Figure 2.11 (d). This correlation was obtained from these suspended sediment concentration measured only during the flood season.

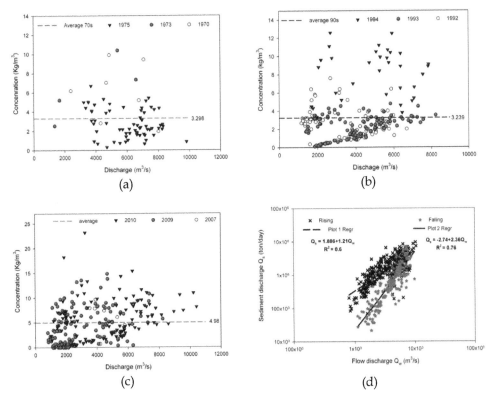

Figure 2.11: Suspended sediment measured at El Deim Station near the Ethiopian Sudanese border in the decades: (a) 1970-1980, (b) 1990-2000, (c) 2000-2010, and (d) Suspended sediment rating curve at El Deim. The measurements were executed during the high flow season (mid June to mid September).

The suspended sediment concentrations measured just downstream of the Roseires Dam outlets is shown in Figure 2.12 (a and b). Figure 2.12 (c) shows the granulometry of suspended sediment at Wad Almahi upstream Roseires Dam and at Wad Al Ais downstream the dam. At Wad Almahi the D_{50} is 18.5μm; at Wad Al Ais the D_{50} is 22µm. This shows that an erosion process happens between these two stations. Silt is the dominant type of sediment in suspension and it represents more than 80% of the samples(Omer, 2011). Sand represents about 15% of the suspended sediment inside the reservoir. The correlation between river flows and the sediment discharges for the rising and falling flood limbs at Wad Al Ais station for the period 1999 – 2009 is depicted in Figure 2.12 (d).

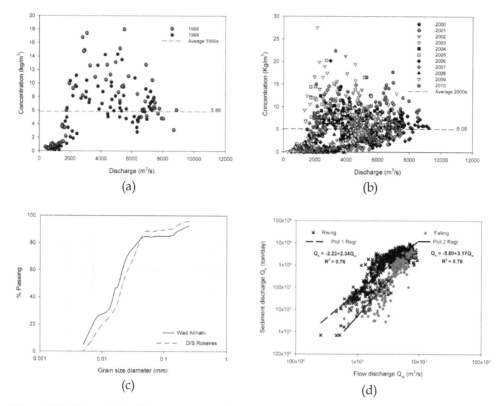

(a)

(b)

(c)

(d)

Figure 2.12: Suspended sediment concentrations measured at Wad Alies, downstream of Roseires Dam during the last two decades: (a) 1990-2000 and (b) 2000-2010 (c)grain size distribution of suspended sediment at Wad Almahi and Wad Alies (D/S Roseires) in 2002(d) Suspended sediment rating curve atWad Alies station.

The correlation between river flows and the sediment discharges at station Downstream Sennar Dam for the period 2002 – 2009 is depicted Figure 2.13 (left).The average monthly sediment concentrations derived from data collected by the Hydraulics Research Center, Ministry of Water Resources and Electricity of Sudan downstream of Sennar Dam integrated with new data from this study as shown in Figure 2.13 (right).

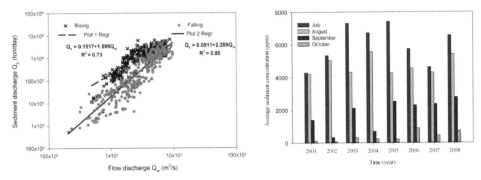

Figure 2.13: Suspended sediment rating curve downstream Sennar (left) and average monthly sediment concentration (right).

Water diverted to Gezira scheme irrigation canals during flood period carries virtually the same suspended sediment concentrations as the Blue Nile River, because in that period the gates of the Roseires Dam are open and the settling rates in the reservoir are low. Data from the Ministry of Irrigation and Water Resources show that between 1933 and 1938 the mean sediment concentration entering the Gezira scheme main canal in August was 700 ppm, whereas the mean sediment concentration in August 1989 was 3800 ppm (Taj Elsir et al., 2001). In the Gezira scheme, for example, more than 70% of the operation and maintenance budgets go for dredging of sediment deposition in irrigation canals and weeds clearance associated with sedimentation (Abdalla, 2006). The sediment loads entering Gezira Scheme through Gezira and Managil at Sennar Dam in the recent years showed increasing trend as depicted in Figure 2.14.

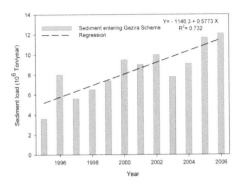

Figure 2.14: the trend of the annual sediment loads entering Gezira Scheme measured at Gezira and Managil canals at Sennar Dam (left) and Gezira main canal downstream of Sennar Dam (right).

2.5   SEDIMENTATION PROBLEMS IN THE LOWER BASIN

2.5.1   Introduction

Sediment transported by water flow could be blessing or threat. On one hand, the deposited sediment renews the soil fertility and lines the channel or canal bed against seepage. On the other hand, it reduces the capacities of reservoirs, inlet channel and irrigation canals. The high sediment load transported by the river during flood season has major influences on the operation of the reservoirs built across the river and the agriculture schemes irrigated from the river. Sediment deposited in reservoirs and irrigation canals reduces the useful life of the reservoirs and canals carrying capacity. Large operational costs are incurred every year in dredging the sediment from reservoirs and canal sediment clearance.

2.5.2   Irrigation canals sedimentation

Sediment normally deposited in irrigation canals under low flow velocity. The deposited sediment is predominantly silt and clay that is inherently fertile and helps to promote rapid aquatic vegetation growth. This further serves to slow down the flow and results in higher rate of deposition.

The Hydraulics Research Center in Sudan and Hydraulics Research Wallingford launched a field data collection programme in 1988 to estimate sediment rates in the Gezira scheme in order to provide recommendations for sediment controls. The results showed that most (97%) of the sediment entering the scheme is silt and clay and about 70% to 80 % of the sediment is entering the irrigation system over a short period from mid July to the end of August (Figure 2.15).

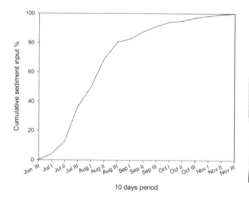

Figure 2.15: the a cumulation of sediment loads entering Gezira Scheme at Sennar Dam (left) and regulating struture (right).

In Gezira scheme, there has been a massive effort to clear sediment from canals, but this has not done with the same rates of sediment settling which resulted in lowering the canals bed and water levels and further difficulties in supplying parts of the scheme. Comparison between the total sediment entering and cleaned in the Gezira Scheme during the period 1988 to 2000 is shown in Figure 2.16.

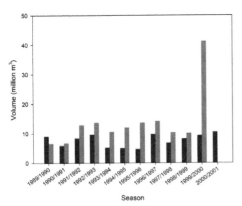 

Figure 2.16: Total sediment entering (blue) and cleaned (red) in the Gezira Scheme during the period 1988 to 2000 (left) and mountains of sediment along the canal banks (right).

2.5.3   Roseires Reservoir sedimentation

Sedimentation inside reservoirs is influenced by many factors, but primarily it is dependent upon the reservoir shape, sediment fall velocity, water flow through the reservoir and reservoir operation (Gottschalk, 1964). Due to the complexity and interaction of many parameters, there are no direct analytical solutions to predict reservoir sedimentation rates. Most of the available methods are therefore either empirical or mathematical and physical models, based on historical data and information from existing reservoirs. The empirical methods are used mostly during the design phase of the reservoir, such as the Area Reduction Method (Annandale, 1987; Borland and Miller, 1958). Other empirical methods are the Brune (1953) method, the Churchill method (1948), the Brown method (1943), Trap efficiency and storage-level approaches.

In Roseires Reservoir, the variations of reservoir storage capacity with time at specific elevation and with elevation in the specific survey year, calculated from the bathymetric surveys of the years 1976, 1981, 1985, 1992, 2005 and 2007 are shown in Figure 2.17. It can be seen that after forty one years of operation (1966-2007), the total capacity of the reservoir (as computed at reduced level R.L = 484 m) have been reduced to 1920.89 million cubic meters having a reduction rate of about 38 million m³/year. However, the reservoir has lost only 13.84 million m³ in the last two years (2005 – 2007) having a reduction rate of about 6.4 million m³/year.

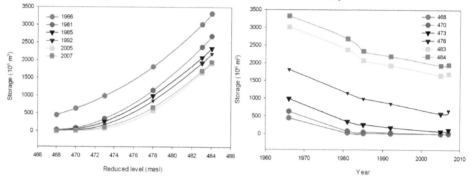

Figure 2.17: The variation of storage with reduce level in the specific survey years (left) and with time at specific reduce level (right).

Trap efficiency

Reservoir trap efficiency is defined as the ratio of deposited sediment to total sediment inflow for a given period within the reservoir economic life. Trap efficiency is influenced by many factors but primarily is dependent upon the sediment fall velocity, flow rate through the reservoir and reservoir operation. The detention-storage time in respect to character of sediment appears to be the most significant controlling factor in most reservoirs (Gottschalk, 1964). Trap efficiency estimates are empirically based upon measured sediment deposits in a large number of reservoirs mainly in USA. Brune (1953) and Churchill methods are the best known ones. Hussein et al. (2005) (Hussein et al., 2005) considered the annual inflow of 50 billion m³ at the 477 m (AD) to estimate the trap efficiency of Roseires Reservoir. The observed and computed trap efficiency values are given in Table (2.4). He estimated the trap efficiency till 1992; we followed the procedure to estimate the trap efficiency in 2005 and 2007.

Table 2.4: Roseires Reservoir Trap Efficiency (%) (after Hussein et al, 2005).

Year of re-survey	1976	1981	1985	1992	2005	2007
T (years)	10	15	19	26	39	41
Observed	45.5	36.0	33.2	28.0	26.1	24
Brune's method	51.0	49.0	46.0	45.0	44.2	43.6
Churchill's method	67.7	66.0	64.4	63.5	62.7	62.9

The deposition rates, however, decreased progressively with time as witnessed from the gradual drop in observed trap efficiency from 45.5% in 1976 to 24% in 2007. This trend was not reflected in the computed trap efficiency values using both Brune's and Churchill's methods which remained fairly constant over the years of observations.

Sedimentation statistical trend (storage-level method)

Empirical methods are normally built on a fair correlation and understanding of physical phenomena through comprehensive data analysis, while mathematical models are normally based on analytical solution of the hydraulic and sedimentation process. The storage-level method adopted in this section, according to (Yevdjevich, 1965) based on the relation between the Storage volume S as a function of both; reservoir elevation H, and time t, and can be approximated by the following function:

$$S = aH^m$$ 2.1

Where: a and m are function of time resulting from sedimentation process, S is the reservoir storage and H is the depth according to a certain level.

When bathymetric survey data are available, a storage-level can be correlated and used to fit the S-H relation to obtain values of a and m parameters in order to be used as a prediction parameters of storage volume through different reservoir's operation time. Storage-Level method is normally simple and direct method to estimate reservoir sedimentation and can obtain a good results if bathymetric survey process conducting regularly with accurate reservoir storage-level schematization data. Nevertheless, the technical and financial constraints limit the opportunity of doing a frequent reservoirs bathymetric survey particularly in developing countries. The results of the historical surveys are given is Table 2.5 including storage capacity (S) in million m³ and depth (H) in m.

Table 2.5: Roseires Reservoir historical surveys including storage capacity and depth.

Reduced level (m)	Depth H (m)	Capacity (million m³)					
		1966	1976	1981	1985	1992	2007
468	2	454	68	36	17.13	11.9	9.05
470	4	638	152	91	60.13	38.97	25.9
473	7	992	444	350	259.2	179.8	113
478	12	1821	1271	1156	992.8	859	660.5
483	17	3024			2082	1937	1701.4
484	18	3329			2337.6	2191.6	1953.8

A relation between S & H for the surveys 1976, 1981, 1985, 1992 and 2007 are shown in Figure 2.18. The fitted values of a and m are summarized in Table 2.6. It should be noticed that H does not represent the stage with mean sea level used as a reference b.

The value of H represents the reservoir depth which is a more meaningful representation (base level 466 m is used).

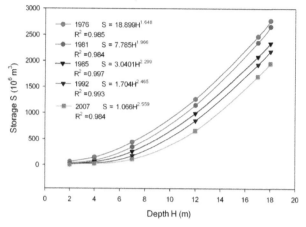

Figure 2.18: Roseires Reservoir, a relation between storage and depth for different surveys.

Table 2.6: The fitted a and m values for different surveys.

Year	Years of operation (t)	a	m	R²
1976	10	18.899	1.6482	0.985
1981	15	7.785	1.966	0.984
1985	19	3.04	2.299	0.997
1992	26	1.704	2.465	0.993
2007	41	1.066	2.559	0.984

Both "a" and "m" varies with time. While "a" is decreasing with time and better expressed in a power, "m" is increasing and has better expressed in a natural logarithmic function (Figure 2.19)

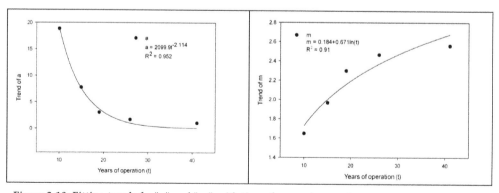

Figure 2.19: Fitting trends for "a" and "m" with time of reservoir operation.

Where t is time in years during which reservoir has been in operation. Using equations (2.2) and (2.3), coefficients "a" and "m" can be predicted for any number of years in which the reservoir has been in operation.

$$m= 0.184+0.671\ln (t)$$ 2.2

$$a= 2099.9t^{-2.114}$$ 2.3

These values can be substituted in equation (2.1) to relate reservoir storage, S, with varying reservoir depth, H. With the possibility of predicting "a" and "m" equation (2.1) becomes a useful tool in estimation of change of storage with both time and reservoir depth. Figure 2.20 shows comparison between the measured and the predicted storage capacity with elevation after 41 years of operation (2007).

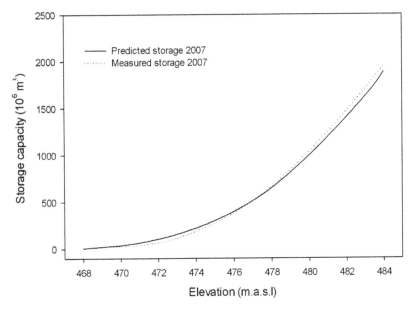

Figure 2.20: Predicted and measured variation of storage capacity with elevation in 2007.

To verify the method of predicting the sediment deposition in reservoirs, the reservoir storage capacity for any year (t) and its preceding year (t-1) was estimated using equations (2.2) and (2.3) to obtain coefficients "a" and "m". Using equation (2.1) the storage capacities St and St-1 in years t and t-1 can be obtained. Then the amount of deposits in year t is obtained by deducting St-1 from St. For the year 1993, after 27 years of operation and maximum operation level of 484 m (H = 18 m), the storage capacity was equal to 2011.69 million m³. Similarly for the preceding year, the storage capacity was calculated and found to be equal to 2024.98 million m³. This means that the amount of sediment deposited inside Roseires Reservoir in the year 1993 was equal to 13.29 million m³ and the amount of sediment entering the reservoir at El

Deim station in the same year was measured to be 125 million tonne (75.76 million m³). Trapping efficiency equal to 17.54 % was obtained by dividing the amount of deposits in the reservoir by the total volume of the sediment entering the reservoir. Alternatively the reservoir trapping efficiency in 1993 is calculated from the measurements conducted in 1993 as shown in Table (2.7) using collected sediment samples upstream and downstream the reservoir and measured discharges. The calculated trapping efficiency is 18.9 %. This finding verifies the method.

Table 2.7 Sediment load estimated at El Deim and downstream Roseires Dam in 1993.

Period	10 day flow (M m³) at El Deim	Sediment concentration at El Deim (ppm)	10 day flow (M m³) at Downstream Roseires	Sediment concentration (ppm) at Downstream Roseires	Sediment load at El Deim (10^6 tonne)	Sediment load at Downstream Roseires (10^6 tonne)
Jun III	1258.0	1956	1311.2	1171	2.46	1.54
Jul I	1663.7	3387	1702.9	2032	5.63	3.46
Jul II	2172.9	3897	2249.8	3193	8.47	7.18
Jul III	4560.6	3941	4528.6	3490	17.97	15.80
Aug I	5546.5	3687	5264.9	3303	20.45	17.39
Aug II	4486.6	3433	4589.8	3090	15.40	14.18
Aug III	5086.2	2948	5243.6	2366	14.99	12.41
Sep I	5480.7	3590	5326.0	2595	19.68	13.82
Sep II	3888.9	2324	3904.0	2177	9.07	8.50
Sep III	3292.6	1734	2357.2	1711	5.71	4.03
Oct I	3595.0	1165	3317.7	833	4.19	2.76
Oct II	2190.0	609	1658.6	312	1.33	0.52
					125.33	101.60

2.5.4   Sennar Reservoir

The variations of reservoir storage capacity with time at specific elevation and with elevation in the specific survey year, calculated from the bathymetric surveys of the years 1981, 1986 and 2008 are shown in Figure 2.21. It can be seen that after eighty two years of operation (1926-2008), the total capacity of the reservoir has been reduced to one third.

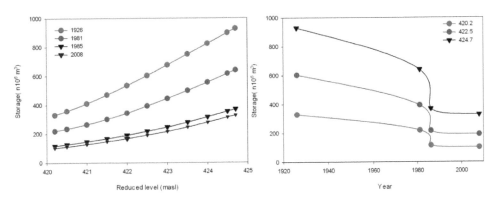

Figure 2.21: The variation of storage with reduce level in the specific survey years(left) and with time at specific reduce level (right).

Trap Efficiency

The sediment balance approach was used to estimate the sediment deposited inside Sennar Reservoir ($Q_{S.D}$), this had been done by estimating the measured sediment load entering the reservoir from the upper boundary at Was El Ais Station ($Q_{s\ in}$), the released sediment load downstream of Sennar Dam ($Q_{s\ out1}$) and sediment load entering Gezira and Managil schemes at Sennar Dam ($Q_{s\ out2}$) as shown in equation 2.4 below:

$$Q_{S.D} = Q_{s\ in} - Q_{s\ out1} - Q_{s\ out2} \qquad\qquad 2.4$$

The results of sediment balance measured during the period from 1995 to 2009 are depicted in Figure 2.22. The estimated trap efficiency during this period has an average value of 17.4%.

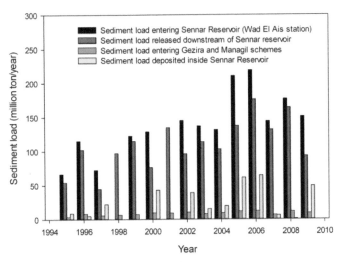

Figure 2.22: Sediment balance along Sennar Reservoir system.

Sedimentation statistical trend (storage-level method)

The results of the historical surveys are given is Table 2.8 including storage capacity (S) in million m³ and depth (H) in m. Reference level for Sennar is taken at 413 m.

Table 2.8: Sennar reservoir historical surveys including storage capacity and depth

Level (m)	Depth H (m)	Capacity (million m³)			
		1926	1981	1985	2008
420.2	7.2	330	220	116	103
420.5	7.5	358	235	126	112
421	8	411	265	145	129
421.5	8.5	471	302	166	147
422	9	537	345	190	168
422.5	9.5	605	395	217	192
423	10	678	446	245	217
423.5	10.5	752	501	278	246
424	11	825	557	315	279
424.5	11.5	900	617	354	314
424.7	11.7	930	640	370	328

A relation between S & H for the surveys 1981, 1985 and 2008 are shown in Figure 2.23.

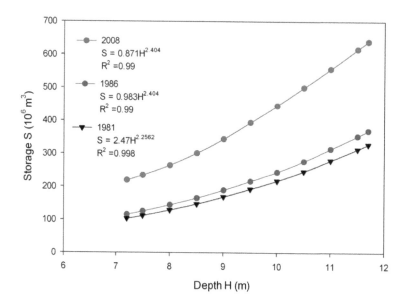

Figure 2.23: Sennar Reservoir, a relation between storage and depth for different surveys.

Since only three surveys in Sennar dam are available, it is not possible here to have good fit as in Roseires Dam for the variation of "a" and "m" with time. But generally as in Roseires "a" decreases while "m" increases with time and both are better expressed in an exponential function (Figure 2.24)

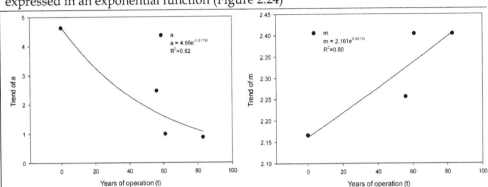

Figure 2.24: Fitting trends for "a" and "m" with time of reservoir operation.

Where t is time in years during which reservoir has been in operation. Using equations (2.5) and (2.6), coefficients "a" and "m" can be predicted for any number of years in which the reservoir has been in operation.

m= 2.161e$^{0.0013t}$ 2.5

a= 4.66e$^{-0.176t}$ 2.6

To verify the method of predicting the sediment deposition inside Sennar Reservoir, the reservoir storage capacity for any year (1996) and its preceding year (1995) were estimated using equations (2.5) and (2.6) to obtain coefficients "a" and "m". Using equation (2.1) the storage capacities S_{1996} and S_{1995} in years 1996 and 1995 can be obtained. For the year 1996, after 71 years of operation and maximum operation level of 424.7 m.a.s.l (H = 11.7 m), the storage capacity was equal to 454.20 million m³. Similarly for the preceding year, the storage capacity was calculated and found to be equal to 458.77 million m³. This means that the amount of sediment deposited inside Sennar Reservoir in the year 1996 was equal to 4.58 million m³ and the amount of sediment entering the reservoir at Wad Al Ais station, about 70 km upstream Sennar Dam body in the same year was measured to be 66.31 million tonne (40.189 million m³). Trapping efficiency equal to 11.4 % was obtained by dividing the amount of deposits in the reservoir by the total volume of the sediment entering the reservoir. Alternatively the reservoir trapping efficiency in 1996 is calculated from the measurements conducted in 1996 as shown in Figure (2.22) using collected sediment samples upstream and downstream the reservoir and the entering and leaving discharges. The trapping efficiency calculated is 12.9 %.

2.6   BLUE NILE BEDFORM

Several samples of bed sediment were collected by the Ministry of Water Resources and Electricity of the Sudan and in the framework of this study by means of a grab sampler, at several locations in 2009. The samples reveal that from Lake Tana to the border between Sudan and Ethiopia the river bed is made by sand, accounting for 80% of the material. In Sudan, the river bed is composed by a mixture of silt and sand. Silt accounts for 50% of the bed material at Khartoum (Figure 2.25).

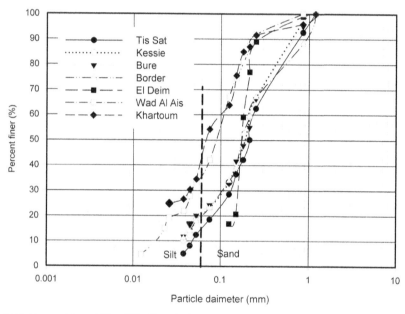

Figure 2.25: Composition of Blue Nile River bed material.

High sediment concentrations have been observed to damp out dunes in river channels at low Froude numbers (Best, 2005; Smith and McLean, 1977). Predicting the occurrence of bedforms is important, because the flow resistance of the river channel is determined by sediment size and form drag. This means that dune formation influences the stage discharge relationship and for this should be included in hydrodynamic models. River dunes are found in nearly all river systems with the river bed consisting of sand, small gravel or mixtures (Julien et al., 2002; Paarlberg et al., 2007; Paarlberg et al., 2010; Wilbers and Ten Brinke, 2003).

The measured average monthly sediment concentrations combined with hydrodynamic information allowed computing the Suspension Number. This is a parameter that characterizes the level of suspension of channel bed material and is defined as:

$$SN = u*/\omega s$$ 2.7

Where *SN* is the Suspension Number, $u*$ is the shear velocity (m/s) and ωs the sediment fall velocity (m/s).

This parameter can be used to infer the presence of dunes on the channel bed, since dunes seem to form for suspension numbers smaller than 2.5 if the Froude number is smaller than 0.32 (Naqshband et al., 2014).

High concentrations of suspended solids were measured during high flow conditions, ranging between 3,000 mg/l and 18,222 mg/l at El Deim and between 3,720 mg/l and 15,044 mg/l just downstream of Roseires Dam. These concentrations results in suspension numbers ranging between 2.8 and 3.5. Given that the corresponding Froude numbers are smaller than 0.32, it is possible to assume that that during high flows dunes do not form in the Sudanese part of the Blue Nile. However, dunes are expected to form at lower flow stages when the suspension number is smaller than 2.5 (Table 2.9).

Table 2.9: Suspension number at El Deim and Wad Al Ais station.

Station	Flow	Q m³/s	W.l m	A m²	h m	U m/s	Fr -	U* m/s	D50 mm	Ws m/s	SN -
El Deim	low	155	488	550	3	0.3	0.05	0.054	0.18	0.0323	1.66
	Medium	960	489	1279	4.6	0.8	0.11	0.067	0.18	0.0323	2.07
	high	6126	493	2974	8.2	2.1	0.23	0.09	0.18	0.0323	2.78
Was Al Ais	low	300	422	758	3.2	0.4	0.07	0.018	0.09	0.0004	2.18
	Medium	944	423	1008	3.6	0.9	0.16	0.019	0.09	0.0004	2.32
	high	5842	426	2175	8.1	2.7	0.3	0.028	0.09	0.0004	3.48

2.7   CONCLUDING REMARKS

Gauging stations along the Blue Nile River and tributaries in Sudan have good record of data but need to be updated. However, in Ethiopia the records of gauged data are limited and new gauging stations need to be installed in particular in the remote areas.

Sediment data in general are limited in the basin in particular the upper basin; comprehensive field campaigns are needed to study the sediment transport along the river system.

Flow and sediment balances are needed to fill the above mentioned knowledge gaps by integrating basin models with available and newly collected data.

Given the problem of lack of data, the work included the following field campaigns:

A. 2009: data collection carried out in Ethiopia. A bathymetric and land survey was carried out to Measure River cross sectional profile, collect soil data from the river banks and bed. The work was carried out during the low flows season (April-June)for many reasons:

(1) During the rainy season (July –October) the flow velocity is very high and the measurement is very difficult (the measured cross section will not be perpendicular to the flow direct).

(2) During the rainy season (July –October) the water depth is relatively big and it is not easy to collect soil samples from the river bed and banks.

(3) The transportation is accessible before the rain season and very difficult to transport during the rainy season in hilly topography.

The difficulties faced the field work can be summarized as follows:

(1) There was no to access to the river in many location which limit the number of measured cross sections.

(2) The sediment concentration sampling from the Blue Nile River system was taken on daily basis but due to budget constrains we were not able to analyze the samples on daily basis.

(3) It was not easy to take samples across border for security reasons which play big role in taking limit soil samples and sediment samples

B. 2010: data collection carried out in Sudan. A bathymetric and land survey was carried out to measure cross sectional profile along the river at many locations. Also soil data were collected from the river banks and bed.

C. 2012: Soil and rock samples collected from the upper basin in Ethiopia from the most eroded areas and rivers bed and banks. In the same year samples from the deposited sediment layers inside Roseires Reservoir were carried out from 4 trenches selected from morphological simulation of the reservoir.

Chapter 3
COMPUTATIONS OF FLOW AND SEDIMENT BALANCES IN THE BLUE NILE RIVER[1]

Summary

This chapter describes the work leading to the quantification of the river flows and sediment loads along the Blue Nile River network through integrating of available data, field survey and modelling tools. Soil and Water Assessment Tool was used to estimate the water flows and sediment loads. Land-use change detection was performed in the most eroded sub basins to link the sediment product with land-use changes.

3.1 BACKGROUND

Estimating the sediment loads along the Blue Nile River is important to assess watershed management programs and to evaluate the sedimentation of the proposed and existing dams along the Blue Nile, since this is necessary to obtain realistic quantifications of the sedimentation rates inside their reservoirs.

The estimation of flow and sediment loads along the entire Blue Nile River network at the sub basin scale through integrating the results of the physics-based hydrological model Soil and Water Assessment Tool (SWAT) (Arnold et al., 1998) with the estimations based on measured data. SWAT has already been applied on a number of Blue Nile sub basins. For instance, it was used to study soil erosion vulnerability in the Lake Tana region (Setegn et al., 2009). The model was further used to predict the impact of climate change on the hydroclimatology of the Lake Tana Basin (Abdo et al., 2009; Setegn et al., 2011). Easton et al. (2011) and White et al. (2011) used the new water balance version of the model (SWAT-WB) to predict flow and soil erosion in the upper Blue Nile. Previous studies focused on the hydrology of

[1]This chapter is based on: ALI Y.S.A., CROSATO A., MOHAMED Y.A., ABDALLA S.H. & WRIGHT N.G. (accepted). Sediment balances in the Blue Nile River Basin. International Journal of Sediment Research, Vol. 29, No 3, pp. 1-13.

the Upper Blue Nile River basin and only a few of them addressed the sediment transport issue (i.e. Steenhuis et al. (2009).

The study area considers the entire Blue Nile River basin including its subdivided in 18 sub basins as described before in chapter 2 (Figure 2.2 b), including the existing reservoirs of Roseires and Sennar. To integrate the available data, this study also included a field campaign in which suspended solids and flow discharges were measured along the Blue Nile and along a number of its major tributaries in Ethiopia. The topography of the study area also was shown in chapter 2 in (Figure 2.2 a).

3.2   MATERIALS AND METHODS

In general, four methods can be applied to assess sediment yield of a river basin. These methods are: 1) calculations from suspended sediment data at gauging stations; 2) estimations of gross soil erosion and sediment delivery ratio; 3) analyses of reservoir sedimentation data; and 4) estimations of sediment transport using sediment transport formulas (Demissie et al., 2004). The first method, however, does not include bed load, whereas the last method does not include wash load. Thus these two methods underestimate the sediment load. This study applies the first two methods. The first one requires the collection of sufficient suspended sediment concentrations and flow discharge data at several locations to develop sediment rating curves (Crowder et al., 2007; Sheridan et al., 2011). The second method estimates the sediment yield from each sub basin using the Soil Water Assessment Tool (Arnold et al., 2012; Arnold et al., 1998). This study also includes a field data campaign in which new data on suspended solids and discharge were collected along the Blue Nile and a number of tributaries.

In this study, data available from the literature were integrated with new data. Time series of daily discharge and suspended sediment concentrations in the period 1970 to 2004 were provided by the Ministry of Water Resources of Ethiopia and Eastern Nile Technical and Regional Office (ENTRO) head quarter in Ethiopia. Daily time series of the discharge measured in the period 1970-2010 and suspended sediment data for the main stations in Sudan were provided by the Hydraulics Research Center in Sudan.

During the 2009 flood season, suspended sediment samples and soil samples were collected in the framework of the study from different locations along the tributaries and the main river, both in Sudan and Ethiopia. The data collection was carried out in Ethiopia because the data are very limited there and in some location there is no data, this due to the fact that the topography is very tough to carry out any measurement. Thirty seven (37) soil samples were taken from different locations in Ethiopia. Fourteen samples were taken from areas affected by soil erosion within the catchment and the remaining samples from the main river and tributaries banks and bed. The suspended sediment concentration was sampled on a daily basis for El Deim station in Sudan during the flood season; June – October and three time a week for Andassa, Gedeb, Kure-Wonez, Dirma, Jemma , Bure and Kessie in Ethiopia . Plastic bottles were used to collect and keep the water samples. The collected samples were analyzed at Addis Ababa University laboratory and Hydraulics research Station laboratory in Wad Medani. Figure 3.1 showed the locations of the collected data.

Figure 3.1: Blue Nile River Basin field campaign 2009: measuring soil samples and suspended sediment concentration. (+ indicate the measuring sites).

Long-term land-use and land-cover changes (LULCC) was detected through the analysis of Land sat imageries of 1973 and 2000 in Jemma, Didessa and South Gojam. Several pre-processing methods were implemented to prepare the land-use maps for classification and change detection including geometric correction, radiometric correction, topographic normalization and temporal normalization.

All scenes supplied by the EROS Data Centre were processed with the Standard Terrain Correction (Level 1T), which provides systematic radiometric and geometric accuracy for the imageries. A hybrid supervised/unsupervised classification approach was carried out to classify the imageries of 1972/1973 (MSS) and 2000 (ETM+). First, Iterative Self-Organizing Data Analysis (ISODATA) clustering was performed to determine the spectral classes. ISODATA is an algorithm frequently used to determine the natural spectral groupings in a dataset for unsupervised classification (Tou and Gonzalez, 1974). Second, ground truth (i.e. reference data) was collected from already classified maps and in-depth interviews were held with local elders to associate the spectral classes with the cover types. Finally, a supervised classification was done using a maximum likelihood algorithm to extract land-use/cover classes from the 1972/1973 and the 2000 imageries.

The accuracy of the classifications was assessed by computing the error matrix that compares the classification result with ground truth information. To assess the accuracy of thematic information derived from 1972/1973 (MSS) and 2000 (ETM+), the "design-based statistical inference" method was employed that provides unbiased map accuracy statistics (Jensen, 2005).Reference data were collected from old maps using stratified random sampling in order to assess the accuracy of the 1973 map. These reference data were assessed using the "confidence-building assessment" method. The confidence-building assessment method involves visual examination of the classified map by knowledgeable local elders to identify any gross errors.

Similarly, in order to assess the accuracy of 2000 map, first unchanged land-cover locations between 2000 were identified by interviewing local elders and with the help of SPOT-5 (5 m resolution) 2007 imagery. Second, several ground truth data regarding land-cover types and their spatial locations were collected from selected sample sites during the field campaign in 2009 using Global Positioning System (GPS). The number of samples required for each class was adjusted based on a proportion class and an inherent variability within each category. As there is no single universally accepted measure of accuracy, overall accuracy and kappa analysis were used to evaluate the accuracy of the classified maps. The overall accuracy was calculated by dividing the number of pixels classified correctly by the total number of pixels.

The post-classification change detection comparison was performed following Jensen's (2005) methodology. This was done to determine changes in land-use/cover between two independently classified maps from images of two different dates.

3.2.1   Data analysis

Sediment rating curve is the representation of the relationship between stream flow discharge and either suspended sediment concentration or suspended sediment load (Walling, 1977; Zhang et al., 2012). The curve is used to derive sediment concentrations from measured flow at different times (Bhutiyani, 2000; Crawford, 1991; Yang et al., 2007). A sediment rating curve, described by Equations 1 and 2, is derived from a logarithmic plot of the stream flow against sediment loads, as a linear regression through the scatter of points.

$$Qs = aQ^b \hspace{10cm} 3.1$$

$$Log(Q_s) = Log(a) + bLog(Q_w) \hspace{8cm} 3.2$$

Where: Q_s is suspended sediment transport (million tonne/day), Q_w is flow (m³/s) and a and b are regression coefficient and exponent, respectively.

Sediment load calculated using the above relation has been reported to underestimate the actual suspended sediment loads (Ferguson, 1986; Walling, 1977; Walling and Webb, 1981). Ferguson (1986) stated that the degree of underestimation can increase with the degree of scatter about the rating curve and can reach 50%. He proposed statistical bias correction factor equal to exp (2.65S^2) to reduce the degree of underestimation by rating curve method:

$$S^2 = \frac{\sum_{i=1}^{n} (Log(C_i) - \overline{Log(C_i)})^2}{n-2} \hspace{7cm} 3.3$$

Where: $S2$ is the variance.

Asselman (2000), found that the rating curves obtained by least squares regression on logarithmic transformed data underestimate the long-term sediment transport rates by 10-50% along the Rhine River and main tributaries. The nonlinear least squares regression was used without correction in the rating curve (Crowder et al., 2007; Demissie et al., 2004). Asselman derived the following optimization procedure (Equation 4):

$$Log\ (Q_s)=Log\ (a)+b(LogQ_w\ )^c \qquad\qquad 3.4$$

Where: a, b, and c are coefficients determined through a regression and optimization procedure.

The above mentioned formats of regression (the normal linear log-log regression, normal log-log regression with correction bias factor and nonlinear least squares regression were used here to derive the sediment yield from measured data. Sediment rating curves were developed at four stations (Kessie Bridge, El Deim, downstream Roseires Dam and downstream Sennar Dam) on the main stream of Blue Nile and on 37 tributaries. The sediment load was derived from the developed rating curve on a daily basis during the rainy season (June-October). The sediment load of each sub basin was estimated by adding the sediment loads from all tributaries within the sub basin. To complete the missing data at certain gauging stations, correlation was made using the available data at the station with the available data from the nearby stations. Example of regression fit to predict missing data at Fincha station is given in Figure 3.2.

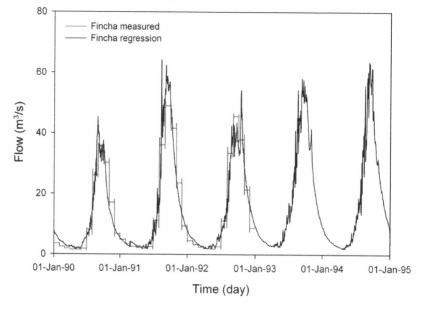

Figure 3.2: Regression fit to predict missing data at Fincha station.

The relative difference between suspended sediment loads calculated from measured suspended sediment concentration and discharge (measured) and predicted suspended loads derived from sediment rating curves (predicted) were used to evaluate the three rating curve approaches to estimate the sediment load. The differences between measured and predicted values may be expressed as a concentration (ppm) as in the case of suspended sediment concentration, as a mass (tonne) in the case of yearly sediment loads, or as a percentage for either measurement. In the latter case, percentage was calculated as follows:

$$\% \ difference = \frac{(predicted \ value - measured \ value)}{measured \ value} \times 100\% \qquad\qquad 3.5$$

3.2.2   SWAT model

The Soil and Water Assessment Tool (SWAT) (www.swat.tamu.edu/software/swat-model)(Winchell et al., 2010) is a physics based model to simulate soil erosion and sediment delivery in a river system. It can be used to estimate the water, sediment and chemical yields in un-gauged basins with varying soil, land-use and management practices over long periods of time, at the basin scale on daily basis (Arnold and Fohrer, 2005; Neitsch et al., 2011).

The model is suitable to simulate a single basin or multiple hydrologically connected basins. The model divides the basin into sub basins based on the size of the basin, the spatial detail of available input data and the amount of detail required. The sub basins should be detailed enough to capture significant topographic variability. The sub- basins are further divided into small portions called "hydrological response units" (HRUs), characterized by uniformity of soil and land-use (Santhi et al., 2001; Stehr et al., 2008). Hydrological processes are simulated in detail for each HRU and then aggregated for the sub basin by a weighted average (Schneiderman et al., 2007). SWAT simulates the water balance of a basin in two phases: the land phase of the hydrological cycle that controls the amount of water, sediment, nutrient and pesticide loads to the main channel and the water or routing phases of the hydrological cycle. This defines the transport of water, sediment, nutrient and pesticide through the channel to the outlet of the sub basin.

The hydrologic cycle of a sub basin simulated by SWAT is based on the following water balance equation:

$$SW_t = SW_0 + \sum_{i=1}^{t}(R_{day} - Q_{surf} - E_a - w_{seep} - Q_{gw}) \qquad\qquad 3.6$$

Where: SW_t is the final water content (mm); SW_0 is the initial water content on day i (mm); t is time (day); R_{day} is the amount of precipitation on day i (mm); Q_{surf} is the amount of runoff on day i (mm); E_a is the amount of evapotranspiration on day i (mm); w_{seep} is the amount of water entering of the vadose zone from the soil profile on day i (mm); and Q_{gw} is the amount of return flow on day i (mm).

Surface runoff occurs if the rate of water input on the ground surface exceeds the rate of infiltration. In SWAT, the surface runoff from daily rainfall is estimated using a modified SCS curve number method (SCS, 1972), which estimates the amount of runoff based on local land use, soil type, and antecedent moisture conditions. SWAT-WB is a modified version of SWAT, which was developed with the explicit goal of accurately modelling surface runoff generation with a physically based soil water balance. SWAT-WB uses the saturation excess mechanism to predict the runoff (Easton et al., 2010; Easton et al., 2011; White et al., 2011)

Sediment transport processes are simulated via soil erosion and sediment export from the hillslopes of the catchment and the sediment processes in the stream channel (Neitsch et al., 2011). The sediment yield from a HRU is calculated using the Modified Universal Equations (MUSLE) which depends on the rainfall runoff energy to entrain and transport sediment (Williams, 1995):

$$sed = 11.8. \left(Q_{surf} \cdot q_{peak} \cdot area_{hru} \right)^{0.56} . K_{USLE} \cdot C_{USLE} \cdot P_{USLE} \cdot LS_{USLE} CFRG \qquad\qquad 3.7$$

Where: sed is the sediment yield on a given day (metric tonne); Q_{surf} is surface runoff volume (mm H_2O / ha); q_{peak} is peak runoff rate (m³/s); $area_{hru}$ is area of HRU (ha); K_{USLE} is the soil erodibility factor (0.013 metric tonne m²hr /(m³-metric tonne cm)); C_{USLE} is the cover and management factor; P_{USLE} is the support practice factor; LS_{USLE} is the topographic factor and $CFRG$ is the coarse fragment factor.

The sediment transported to the main channel in a given day depends on the amount of sediment load generated in the HRU in that day, the sediment stored or lagged from previous day, the surface runoff lag coefficient and the time of concentration for the HRU. The sediment discharged to the main channel is calculated based in the following equation (Neitsch et al., 2011):

$$sed = \left(sed' + sed_{stor,i-1} \right). \left[1 - exp \left(\frac{-surf_{lag}}{t_{conc}} \right) \right] \qquad\qquad 3.8$$

Where: sed is the amount of sediment discharged to the main channel on a given day (metric tonne); sed' is the amount of sediment load generated in the HRU on a given day (metric tonne); $sed_{stor,i-1}$ is the sediment stored or lagged in the HRU from previous day (metric tonne); $surf_{lag}$ is the surface runoff lag coefficient and t_{conc} is the time of concentration for the HRU (hrs).

There are several statistical measures that can be used to evaluate the performance of models, among which the Nash-Sutcliffe model efficiency (NSE) parameter. This indicates how well the plot of observed versus simulated data fits the 1:1 line. The following equation is used to calculate NSE:

$$NSE = 1 - \frac{\sum_{i=1}^{n}(y_i - x_i)^2}{\sum_{i=1}^{n}(y_i - \bar{y})^2}$$

3.9

Where: x is the simulated value and y is the actual value.

3.2.3   Model development

The setup of a SWAT model requires data as topographic maps (DEM), land-use maps, soil maps and meteorological information, such as daily precipitation, maximum and minimum temperatures, solar radiation, relative humidity and wind speed. Daily data of precipitation, maximum and minimum temperature, daily solar radiation was provided by the Methodical Organization and the Ministry of Water Resources of Sudan, as well as from the Eastern Nile Technical and regional Office and the Ministry of Water Resources of Ethiopia. The DEM provided in chapter 2 in Figure 2.2 a was obtained from the Shuttle Radar Topographic Mission (SRTM), available at http://srtm.csi.cgiar.org; with a resolution of 90 m. DEM is used to analyze the drainage pattern of the watershed, slope, stream length, width of channel within the watershed. The soil map was obtained from the Food and Agriculture Organization of the United Nations (http://www.fao.org/ag/agl/agll/dsmw.stm) (FAO, 1995). They provide about 5000 soil types at a spatial resolution of 10 kilometres with soil properties for two layers (0-30 cm and 30-100 cm depth). The most dominant soils in the basin are Vertisols, Nitisols, Luvisols and Leptosols among, others. The soil map of the Blue Nile River Basin is given in Figure 3.3. The Land-use map was obtained from the Global Land-cover Characterization (GLCC) database   of   the   United   States   Geological   Survey   (USGS) (http://edcsns17.cr.usgs.gov/glcc/glcc.html). The map has a spatial resolution of 1 km and 24 land-use classes (Figure 3.4).

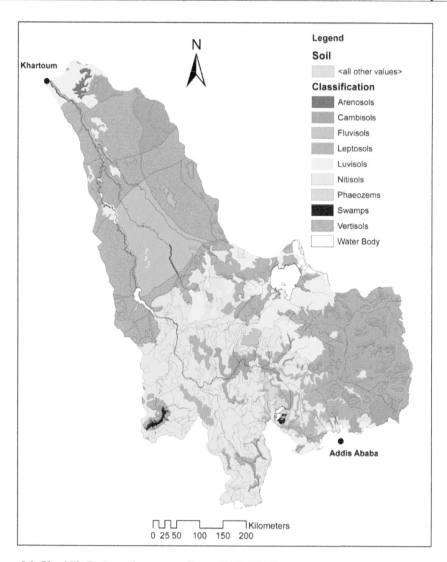

Figure 3.3: Blue Nile Basin- soil map according to FAO (1995).

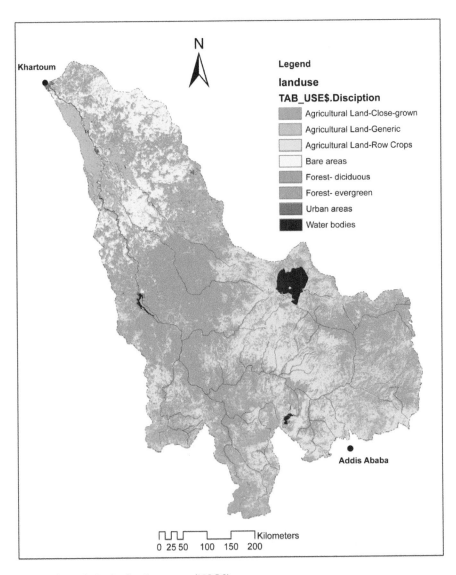

Figure 3.4: Blue Nile Basin- land-use map (USGS).

3.2.4   Model calibration and validation

Performing sensitivity analyses is helpful to identify the most influential parameters governing model results (van Griensven et al., 2002; van Griensven et al., 2006). Furthermore, the results of the sensitivity analyses allow determining the effects of varying the values of these parameters to get an insight on the uncertainty of the results.

Sensitivity analyses were therefore carried out for forty two parameters selected from the SWAT documentation (Neitsch et al., 2011) and based on the results of previous studies within the basin (Betrie et al., 2011; Setegn et al., 2011; Setegn et al., 2009)to identify the most sensitive parameters that affect the water flow and the sediment yield.

The most sensitive parameters resulted from the sensitivity analysis were adjusted until the output from the model gives an acceptable agreement with the actual measurement. The parameters were adjusted manually and then automatically using sequential uncertainty fitting (SUFI-2) algorithm (Abbaspour et al., 2004; Abbaspour et al., 2007). The SUFI-2 algorithm is linked to the SWAT model using SWAT Calibration and Uncertainty Procedures (SWAT-CUP) (Abbaspour et al., 2009; Arnold et al., 2012; Rouholahnejad et al., 2012). Two measures were used to assess the goodness of calibration and uncertainty analysis: (a) the percentage of data bracketed by the 95% prediction uncertainty (95PPU) (P factor) and (b) the ratio of average thickness of the 95PPU band to the standard deviation of the corresponding measured variable (D factor). Ideally, the P factor should tend towards 1 with a D factor close to zero (Rostamian et al., 2008).

3.3   MODEL APPLICATION

3.3.1   Results of sensitivity analyses

The SWAT model was evaluated by performing sensitivity analyses, calibration and validation using the SUFI-2 (sequential uncertainty fitting version 2) algorithm (Abbaspour et al., 2004), which is a semi-automated inverse modelling procedure for a combined calibration-uncertainty analysis.

The objective function selected during the sensitivity analyses was the sum of the squared errors between observed and simulated values. The rank of the most sensitive parameters for river flow and sediment load from the most sensitive to least sensitive are given in Table 3.1. The most sensitive parameters for river flow and sediment loads were further used for calibration and uncertainty analysis of the model via SWAT-CUP. The most sensitive flow parameters resulted were the surface runoff parameters, including the initial moisture condition II curve number (CN2), which is a function of soil's permeability and land use. Moreover, parameters such as maximum canopy storage (CANMX) and average channel effective hydraulic conductivity (CH_K2) were also ranked as highly sensitive to the stream flow. The plant canopy can affect the surface runoff and infiltration rate either by reducing the erosive energy of rain droplet and traps a portion of the rainfall within the canopy or contributes more evapotranspiration from surface and soil water which was supposed to flow out through surface runoff and base flow (Neitsch et al., 2011). High sensitivity of Ch_K2 shows that the channel effective hydraulic conductivity governs the loss of stream flow through leaching at the watercourse to the ground water system.

The most sensitive parameters for sediment transport prediction were the channel roughness, expresses as Manning n value (Ch_N2), the linear re-entrainment parameter for channel sediment routing (Spcon) and the Channel hydraulic conductivity (Ch_K2). The sediment yield from the basin was very sensitive to the value of SPCON, as it affects deposition in the channel.

Table 3.1: Rank of sensitive parameters resulted from Sensitivity analysis of SWAT model.

| | | Rank | |
Parameter	Description	Flow (m³/s)	Sediment (ppm)
Cn2	Moisture condition II curve number	1	5
Canmx	Maximum canapoy storage (mm H2O)	2	10
Ch_K2	Channel hydraulic conductivity (mm/hr)	3	3
Surlag	Surface runoff lag time (days)	4	7
Sol_Awc	Available water capacity (mm H2O/mm soil)	5	13
Blai	Maximum potential leaf area index	6	17
Ch_Erod	Channel erodibility factor	7	42
Ch_N2	Channel manning n value	8	1
Sol_Z	Depth to bottom of soil layer (mm)	9	14
Esco	Soil evaporation compensation factor	10	12
Alpha_Bf	Base flow alpha factor (days)	11	8
Rchrg_Dp	Deep aquiver percolation fraction	12	42
Gw_Revap	Ground water revap coefficient	13	23
Epco	Plant uptake compensation factor	14	19
Sol_K	Saturated hydraulic conductivity (mm/hr)	15	21
Slsubbsn	Average slope length (m)	16	15
Gwqmn	Threshold depth of water in the shallow aquifer for flow	17	20
Biomix	biological mixing efficiency	19	22
Usle_C	Minimum USLE cover factor	20	18
Usle_P	USLE management support	21	6
Spcon	Linear re-entrainment parameter for channel sediment	42	2
Spexp	Expon. re-entrainment parameter for channel sediment	42	4
Revapmn	Thresh hold water depth in the shallow aquifer for "revap"	42	11
Slope	Slope of water shed	42	9
Gw_Delay	ground water delay (day)	42	16
Sol_Alb	Moist soil albedo	42	24

3.3.2   Results of model calibration and validation

SWAT-CUP was applied to the Blue Nile River basin to compare the computed flow and sediment loads with the data measured at some stations. The model was calibrated against the historical data collected at El Deim and Kessie Bridge in the period 1992-1996 and validated against the data collected in the period 2001-2007 at El Deim and in the period 2001- 2005 at Kessie Bridge. The results of calibration are given in Figure 3.5. The model was able to simulate the peak flows and sediment loads quite well. Similarly, the model could capture dry period characteristics as well.

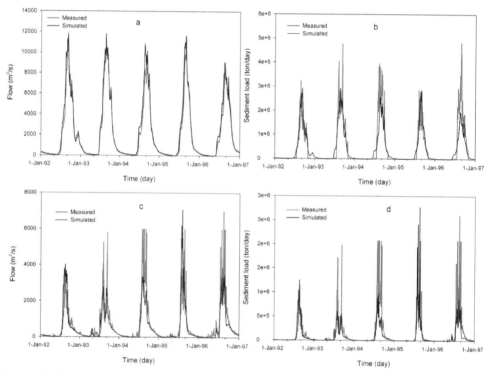

Figure 3.5: SWAT model calibration at El Deim station (a) flow (b) sediment load and at Kessie Bridge station (c) flow and (d) sediment load.

The results of model validation are given in Figure 3.6.The Nash-Sutcliffe Efficiency (NSE) during calibration resulted in 0.95 and 0.75 for flow and sediment load, respectively at El Deim station. At Kessie Bridge the model performance was less good than in El Deim, with Nash-Sutcliffe Efficiency of 0.71 and 0.53 for flow and sediment load, respectively. For validation, the Nash-Sutcliffe Efficiency (NSE) was 0.95 and 0.68 for flow and sediment, respectively at El Deim and slightly lower at

Kessie Bridge with Nash-Sutcliffe Efficiency of 0.91 and 0.52 for flow and sediment respectively. The results obtained at El Deim Station are better than Kessie Bridge Station because El Deim Station is located at very stable site with long measurement records. However, Kessie Bridge is located near the bridge that can affect the measurements.

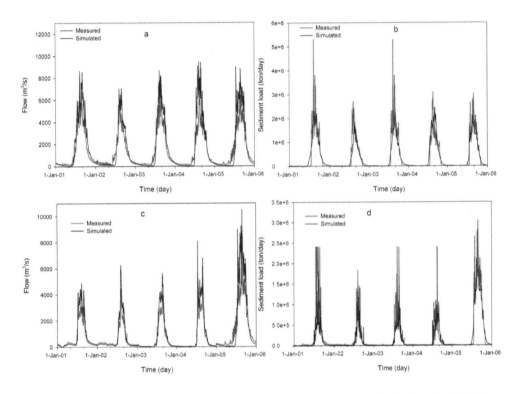

Figure 3.6: SWAT model validation at El Deim station (a) flow (b) sediment load and at Kessie Bridge station (c) flow and (d) sediment load.

3.3.3   Result of water balance

The model was applied to estimate water flows from the un-gauged basins such as Jemma and some parts of Jemma basin. The application of the model with the available data allowed deriving a complete water balance of the basin as depicted in Figure 3.7. The long term annual flow measured at El Deim Station was found to be 45.7 billion m³/year. However, the sum of flow contribution from sub basins upstream El Deim was resulted in 44.96 billion m³/year.

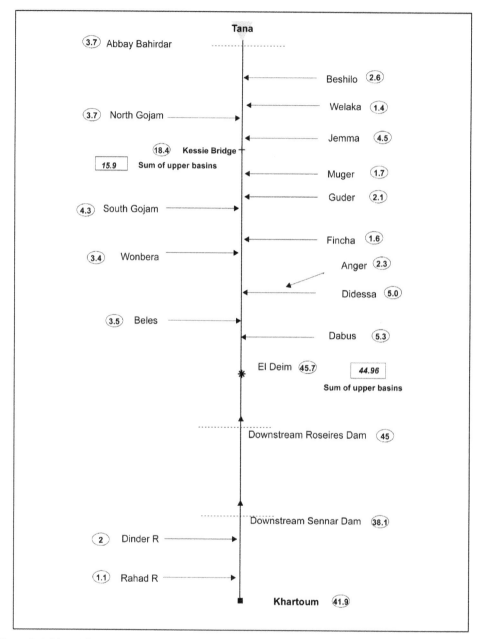

Figure 3.7: Blue Nile River Basin network annual averaged (1980-2004) flow contribution in billion m3/year.

3.3.4   Assessment of sediment loads

Example of developed rating curves at Kessie Bridge, El Deim; downstream of Roseires Dam and downstream of Sennar Dam are given in Figure 3.8. The rating curves were developed using the linear log-log regression and the nonlinear log-log regression, while the statistical bias correction factor was estimated from the measured sediment concentration at each station. It should be noted that limited data are available at Kessie Bridge including the measured data in 2009.

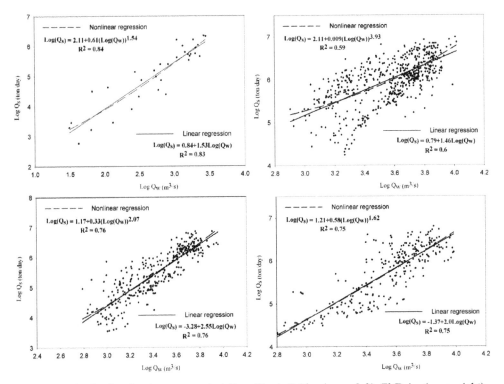

Figure 3.8: The developed rating curves at stations Kessie Bridge (upper left), El Deim (upper right), downstream of Roseires Dam (lower left) and downstream of Sennar Dam (lower right).

The results of developed rating curve at five stations in the main river and thirty five stations in the tributaries are summaries in Table 3.2.

Table 3.2: Developed rating curves along the Blue Nile River and tributaries.

River name	linear regression Log (Qs) = a+bLog(Q)			Corr. Factor	Non linear regression Log (Qs) = a+bLog(Q)ᶜ		
	a	b	r²	εC.F	a	b	c
Main Beles	0.79	1.62	0.79	1.50	0.28	2.20	0.79
Gelgel Beles	2.44	2.00	0.83	1.75	0.89	1.53	1.19
Jemma	10.05	1.56	0.64	2.23	2.98	0.62	1.02
Dabus	9.11	1.28	0.84	2.69	0.86	1.29	1.23
Didessa	4.16	1.31	0.96	1.15	1.02	0.84	1.40
Anger	8.64	1.12	0.94	1.15	1.02	0.84	1.40
Fincha	6.13	1.88	0.65	1.86	0.41	2.29	0.63
Guder	5.85	1.16	0.90	1.40	-3.12	5.29	0.08
Muger	33.70	1.69	0.87	1.54	-0.08	3.62	0.39
Jemma (Jemma)	7.01	1.63	0.83	5.53	0.40	2.06	0.85
Beressa (Jemma)	21.00	1.20	0.73	3.60	0.83	1.90	0.66
Wizer (Jemma)	1.17	1.45	0.98	1.25	0.35	3.31	1.00
Aliltu (Jemma)	14.49	1.22	0.91	2.21	1.58	0.84	1.18
Tigdar (North Gojam)	56.26	1.46	0.87	3.48	1.19	2.07	0.70
Andassa (NG)	7.16	2.00	0.82	2.14	-44.38	47.00	0.07
Sedie (North Gojam)	42.39	1.38	0.76	1.49	0.17	1.24	1.31
Shina (North Gojam)	12.91	1.22	0.93	3.70	1.15	12.26	3.93
Mendel (North Gojam)	103.36	1.91	0.89	3.88	1.83	1.64	0.61
Azuari (North Gojam)	31.25	1.30	0.65	3.49	1.86	0.90	1.12
Muga (North Gojam)	4.58	2.12	0.86	2.19	1.53	1.16	1.86
Suha (North Gojam)	45.55	1.43	0.90	4.54	2.07	0.88	2.32
Welaka (selegie)	21.62	1.18	0.66	3.60	1.03	4.19	2.41
Bahridar (Tana)	6.23	1.36	0.79	2.84	1.40	0.90	1.10
Ardie (South Gojam)	14.78	1.18	0.72	4.91	1.12	0.96	1.68
Dura (South Gojam)	146.22	0.95	0.94	1.68	3.16	0.10	3.31
Buchiksi (South Gojam)	1.26	1.16	0.98	1.03	0.10	1.16	1.00
Lower Fettam(South Gojam)	15.35	1.40	0.73	1.74	-1.85	3.25	2.65
Birr ((South Gojam)	53.40	1.44	0.70	2.23	2.38	0.68	1.84
Yeda (South Gojam)	113.83	1.12	0.83	1.33	2.13	5.79	1.78
Chamoga (South Gojam)	90.06	1.29	0.89	1.99	1.59	1.72	0.86
Jedeb (South Gojam)	88.56	1.11	0.69	2.79	0.72	2.21	0.66
Temcha (South Gojam)	26.92	1.14	0.80	2.16	1.18	1.63	0.49
Bogena (South Gojam)	33.08	1.46	0.98	1.70	-1.41	4.51	0.47
Gulda (South Gojam)	11.03	1.47	0.64	1.25	-17.71	19.75	0.16
Kessie	0.84	1.53	0.83	1.23	2.11	0.61	1.54
El Deim	0.75	1.46	0.45	1.30	4.57	0.01	3.93
Downstream Roseires Dam	-3.28	2.55	0.76	1.21	1.17	0.33	2.07
Downstream Sennar Dam	-1.37	2.00	0.75	1.24	1.21	0.58	1.62
Dinder	2.14	0.82	0.51	1.33	-41.63	46.66	0.01
Rahad	2.53	0.72	0.41	1.54	-14.94	17.95	0.08

The results of the sediment loads were obtained on yearly basis. The long-term annual average sediment loads obtained from the normal linear log-log regression, normal log-log regression with correction bias factor and non-linear log-log regression are shown in Figure 3.9, Figure 3.10 and Figure 3.11.

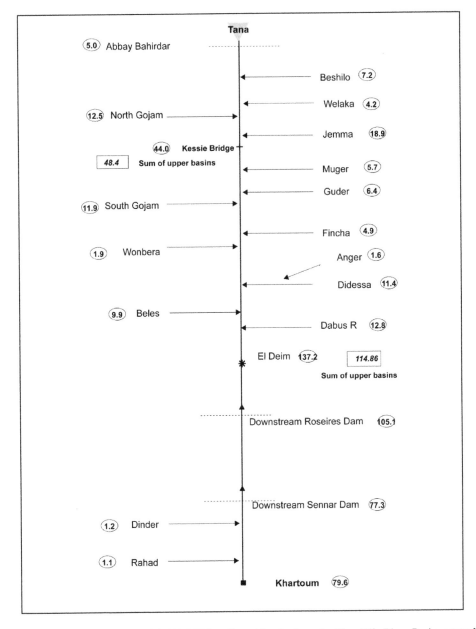

Figure 3.9: Long-term averaged (1980-2004) sediment loads along the Blue Nile River Basin network in million tonne/year estimated from linear regression rating curves.

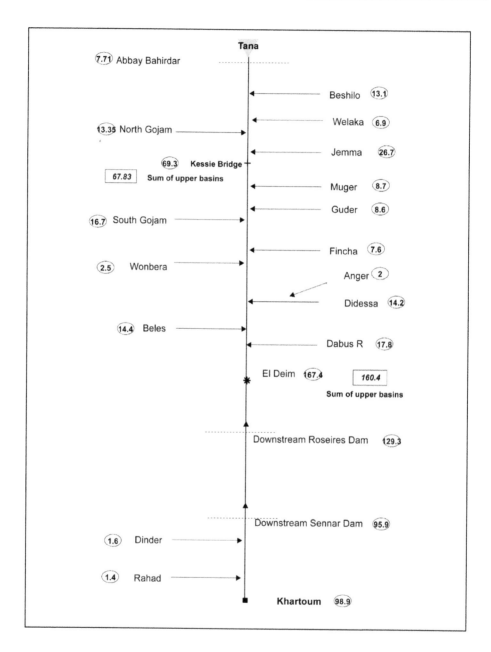

Figure 3.10: Long-term averaged (1980-2004) sediment loads along the Blue Nile River Basin network in million tonne/year estimated from the linear regression rating curves with correction factor.

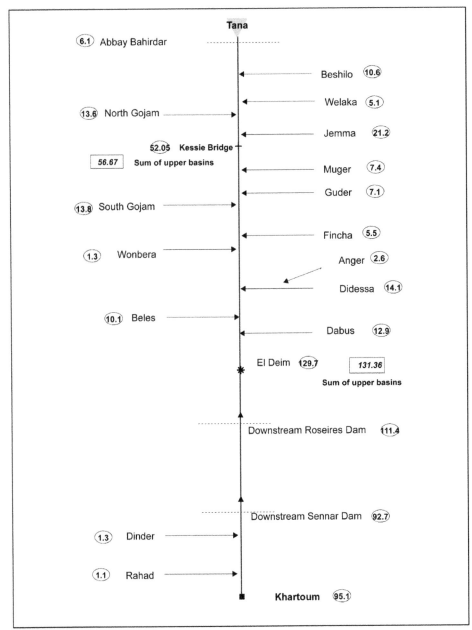

Figure 3.11: Long-term average (1980-2004) sediment loads along Blue Nile River Basin network in million tonne/year estimated from the nonlinear regression rating curves.

The long-term annual average sediment load at Kessie Bridge was found to be 44.8, 69.3 and 52.05 million tonne using the three above mentioned approaches, respectively, while the contribution from the basins upstream Kessie Bridge resulted in 48.4, 67.83 and 56.67 million tonne, respectively.

The same comparison was performed at the measuring station of El Deim, at the border between Sudan and Ethiopia. The annual average sediment load was found to be 137.2, 167.4 and 129.7 million tonne using the three above mentioned methods, respectively, while the sums of the upstream basins (without Kessie station, since it is on the main river stream) resulted in 114.86, 160.36 and 131.36 million tonne, respectively.

The actual suspended sediment load measured in the frame work of this study for the Blue Nile River at El Deim station, based on daily samples collected in the flood season of 2009 was 141.6 million tonne. On the other hand, the estimated suspended loads from rating curve based on the same daily values was 94.62 million tonne, a 33% underestimate when using liner regression approach, 123 million tonne, a 13% underestimate when using liner regression approach with correction factor and 138.7 million tonne, a 2.5% underestimate when using nonlinear regression approach.

The results show that most of the suspended sediment is coming Jemma sub basin, the long term average annual sediment load has been found to be 21.2 million tonne/year (13.4 tonne/hectare/year) and the minimum contribution of suspended sediment load in the basin is coming from Anger sub basin as found to be has been 1.6 million tonne/year (2 tonne/hectare/year).

SWAT was also applied to estimate sediment loads at Kessie Bridge and El Deim and its results compared with data. The long-term average annual sediment loads obtained from SWAT model at Kessie and El Deim stations was found to be 37.5 and 118 million tonne respectively. Whereas, suspended sediment measured in the years 1992 and 1993 at El Deim Station was reported to be 128 million tonne per year, whereas Easton et al (2010) using SWAT model in the upper Blue Nile Basin predicted 121 million tonne/year over the year 1992 and 1993.

3.4   LAND-USE CHANGE DETECTION

The sediment computations showed that the most affected areas by erosion are Jemma, Didessa and South Gojam respectively. It is believed that one reasons of the increasing of erosion in the upper basin is due to land-use changes/land-cover changes (LULCC)(Gebremicael et al., 2013). The LULCC was detected after 28 years to verify the sediment contributions from these sub basins. The analysis of Land sat imageries of 1973 and 2000 in Jemma, Didessa and South Gojam showed eleven land-use categories as described in Table 3.3.

Table 3.3: General description of Land-use categories in Jemma, Didessa and South Gojam sub basins.

LC_ID	Description	General description
1	Rainfed crop land	Area covered with temporary crops grown by rainfall
2	Grass land	Areas in which grasses are dominant
3	Wooded grass land	Lands with herbaceous and tree canopy cover of 10–40%
4	Wood land	A single storey trees and exceed 5 m in height
5	Shrubs and bushes	Low woody plant (<2 m) with multiple stems
6	Natural Forest	Evergreen/deciduous broadleaf forest
7	Water body	Area covered with lakes, reservoirs, and ponds
8	Afro alpine vegetation	High altitude herbaceous and Erica/Hypericum forest
9	Barren land	Areas with little or no vegetation consisting of exposed soil/rocks
10	Irrigated crop land	Area covered with temporary crops grown by irrigation
11	Plantation Forest	Plantation of Eucalyptus globules and Cupresus spp.

3.4.1   Jemma Sub basin

The supervised classification of Jemma sub basin was done using a maximum likelihood algorithm to extract nine land-use/cover classes from the 1972/1973 imageries and eight classes from the 2000 imageries. Comparison of areas and rates of change of the land-cover categories between 1973 and 2000 are presented in Table 3.4.

Table 3.4: Comparison of areas and rates of change of the nine land-cover (L.C) categories between 1973 and 2000.

Description	1973 L.C (1000 Km²)	2000 L.C (1000 Km²)	1973 L.C (%)	2000 L.C (%)	Rate of change in L.C (1000 km²/year)
Rainfed crop	4.78	11.15	30.53	71.23	0.2360
Grass land	2.01	1.00	12.85	6.35	-0.0377
Wooded grass	7.17	2.67	45.78	17.04	-0.1667
Wood land	0.59	0.19	3.76	1.21	-0.0148
Shrubs and	0.95	0.52	6.06	3.33	-0.0158
Natural Forest	0.05	0.0003	0.31	0.00	-0.0018
Afro alpine	1E-6	0.00	0.0006	0.00	0.0000
Barren land	0.11	0.21	0.71	1.33	0.0036
Plantation	0.00	0.04	0.00	0.27	0.0016

In 1973, Jemma sub basin (Figure 3.12 left) was dominated by wooded grassland (45.8%), followed by rainfed cropland (30.5%) and grass land (12.8%). In 2000, the basin (Figure 3.12 right) was dominated by rainfed cropland (71.2%), followed by wooded grass land (17%) and grassland (6.4%).

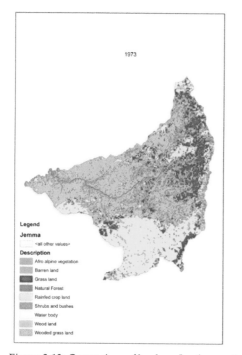
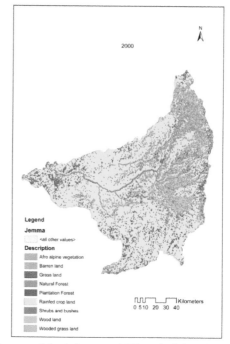

Figure 3.12: Comparison of land use/land-cover in Jemma sub basin between 1973 to the left and 2000 to the right.

The areal coverage of rainfed cropland showed a growth rate of 0.236 km²/year. However, wooded grassland, wood land, shrubs and bushes, natural forest, afro-alpine vegetation in the sub basin showed a decline rate range between -0.0018 and -0.04667 km²/year. Comparison between the land-use / land-cover in 1973 and 2000 is described in and Figure 3.13.

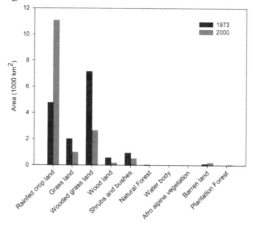

Figure 3.13: Comparison land use/land-cover area in Jemma sub basin between 1973 and 2000.

The land use change detection was extended to recent year with some focus on the upper and middle part of the sub basin to determine the changes till recent years. Satellite image of 2010 was used in addition to the 1973 and 2000 images to detect the land-use changes as described in Figure 3.14. The results showed significant increase in the cultivated area (21%) in 1973 to 65% in 2000 and further to 79% in 2010. As a consequence, woodland, wooded grassland and grassland decreased from 70% of the total area in 1973 to 29% in 2000 15% in 2010.

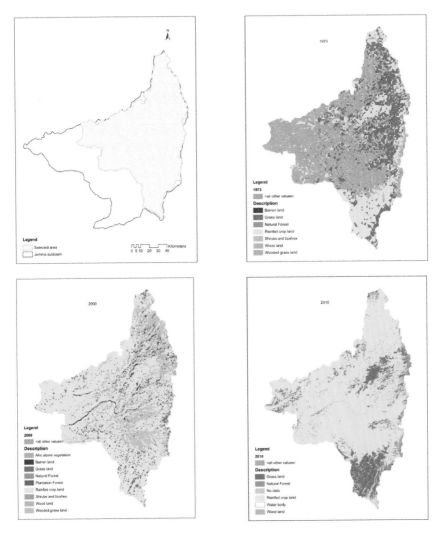

Figure 3.14: Comparison of land use/land-cover in the upper and middle Jemma sub basin between 1973 (upper right), 2000 (lower left) and 2010 (lower right).

The observed land-use change pattern in Jemma sub basin, namely the deforestation of natural woody vegetation and the expansion of cultivated land, is consistent with the results of higher sediment production in the basin. It is therefore probable that the observed increasing sediment load from the Upper Blue Nile basin are caused by land-use change over the basin, and in particular by the conversion of the natural vegetation cover into the agricultural crop land. These results agree with other local level studies in Jemma sub basin (Tekle and Hedlund, 2000; Zeleke and Hurni, 2001).

3.4.2   Didessa Sub basin

In Didessa sub basin, six and seven land-use categories were identified in 1973 and 2000 respectively as shown in Table (3.5).

Table 3.5: Comparison of areas and rates of change of the seven land-cover (L.C)categories between 1973 and 2000.

Description	1973 L.C (1000 Km²)	2000 L.C (1000 Km²)	1973 L.C (%)	2000 L.C (%)	Rate of change in L.C(1000 Km²/year)
Rainfed crop	3.32	9.21	17.01	47.25	0.2180
Grass land	0.25	1.71	1.26	8.75	0.0541
Wooded	3.68	0.36	18.81	1.85	-0.1228
Wood land	8.73	6.36	44.69	32.65	-0.0877
Shrubs and	0.92	0.98	4.70	5.05	0.0025
Natural	2.64	0.86	13.53	4.41	-0.0661
Water body		0.01	0.00	0.04	0.0003

The land use in 1973 (Figure 3.15 left) was dominated by woodland (44.7%), followed by wooded grassland (18.8%), rainfed cropland (17%) and natural forest (13.5%). As a result of land-use changes in2000 (Figure 3.15 right), the basin was dominated by rainfed cropland (47.3%), followed by wood land (32.7%) and grassland (8.8%).

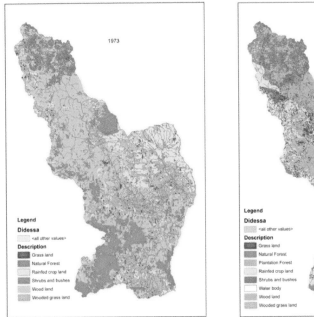
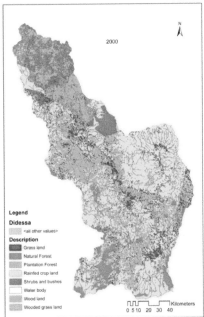

Figure 3.15: Comparison of land use/land-cover in Didessa sub basin between 1973 to the left and 2000 to the right.

In this period, the rainfed cropland showed a growth rate of 0.218 km²/year. However, declining in wooded grassland, wood land and natural forest was occurred by a rate ranging between -0.06 and -0.12 km²/year as described and Figure (3.16)

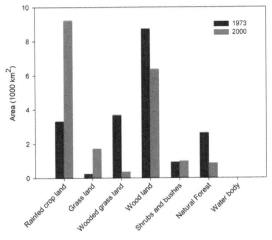

Figure 3.16: Comparison land use/land-cover area in Didessa sub basin between 1973 and 2000.

The land-use change detection was extended to recent year with some focus on part of the sub basin to determine the changes in Didessa Arjo area (Figure 3.17). Satellite image of 2010 was used in addition to the 1973 and 2000 images to detect the land-use changes in Didessa Arjo area where a large agriculture project is recently implemented. The results showed significant increase in the cultivated area (21%) in 1973 to 53% in 2000 and further to 71% in 2010. As a consequence, woodland, wooded grassland and grassland decreased from 73% of the total area in 1973 to 47% in 2000 17% in 2010.

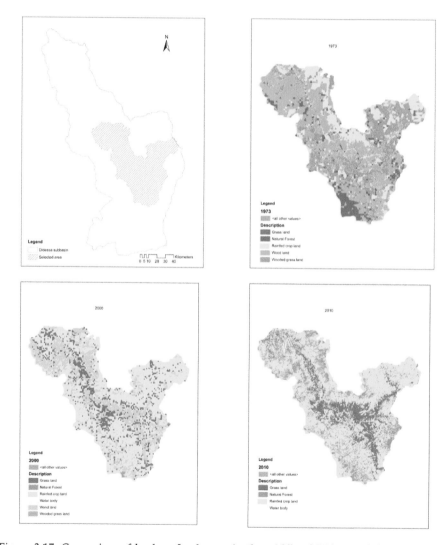

Figure 3.17: Comparison of land use/land-cover in the middle of Didessa sub basin between 1973 (upper right), 2000 (lower left) and 2010 (lower right).

The land-use change pattern in Didessa sub basin, namely the deforestation of natural woody vegetation took place immediately after the 1983/84 drought and at the time government change (1991).However, at the mid of 1990s the government started controlling measures to minimize the rate of deforestation (Sima, 2011). The sub basin is subjected to expansion in cultivated land such as Didessa Arjo agricultural Scheme among others (Johnston and McCartney, 2010).

3.4.3   South Gojam Sub basin

In this sub basin, seven and ten land-use categories were identified in 1973 and 2000 respectively as shown in Table (3.6).

Table 3.6: Comparison of areas and rates of change of the seven land-cover categories between 1973 and 2000.

Description	1973 L.C (1000 Km²)	2000 L.C (1000 Km²)	1973 Land-cover (%)	2000 Land-cover (%)	Rate of change in L.C (1000 km²/year)
Rainfed crop	8.30	10.31	49.50	61.49	0.0744
Grass land	0.80	2.91	4.75	17.38	0.0784
Wooded	3.25	0.59	19.37	3.50	-0.0985
Wood land	3.31	1.71	19.78	10.19	-0.0595
Shrubs and	0.74	0.72	4.42	4.27	-0.0009
Natural	0.35	0.25	2.07	1.49	-0.0036
Water body	0.00	0.01	0.00	0.05	0.0003
Afro alpine	0.018	0.02	0.11	0.09	-0.0001
Barren land	0.00	0.0038	0.00	0.02	0.0001
Plantation	0.00	0.25	0.00	1.51	0.0094

The land use in 1973 (Figure 3.18 left) was dominated by rainfed cropland (49.5 %), followed by woodland (19.78%) and wooded grassland (19.37%). As a result of resources exploitation, land-use changes in2000 (Figure 3.18 right) showed increasing of the rainfed cropland and grassland to 61.49% and 17.38% respectively by a rate of 0.074 and 0.099 km²/year respectively. However, areas of wood land (10.19%) and wooded grassland (3.5%) showed a declining trend by a rate of -0.06 and -0.099 km²/year respectively as described in Figure (3.19).

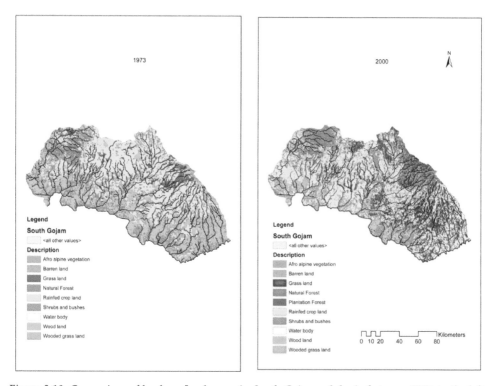

Figure 3.18: Comparison of land use/land-cover in South Gojam sub basin between 1973 to the left and 2000 to the right.

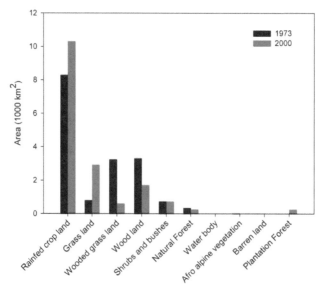

Figure 3.19: Comparison land use/land-cover area in South Gojam sub basin between 1973 and 2000.

The land-use change detection was extended to recent year with some focus on the Eastern of the sub basin where severe erosion takes place in Choke Mountain. Satellite image of 2010 was used in addition to the 1973 and 2000 images to detect the land-use changes as described in Figure 3.20. The results showed increasing trend in the cultivated area (48%) in 1973 to 53% in 2000 and further to 63% in 2010. As a consequence, woodland, wooded grassland and grassland decreased from 47% of the total area in 1973 to 40% in 2000 25% in 2010.

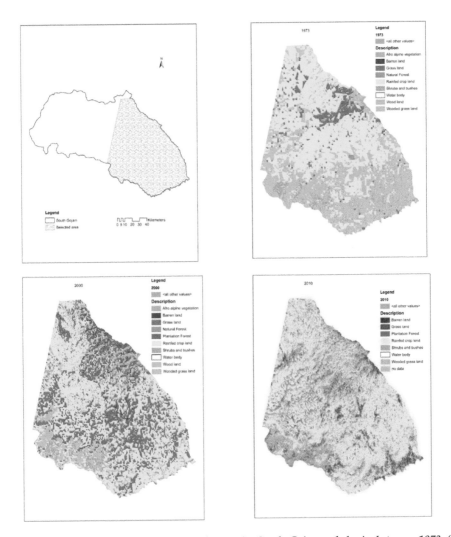

Figure 3.20: Comparison of land use/land-cover in South Gojam sub basin between 1973 (upper right), 2000 (lower left) and 2010 (lower right).

The land-use change pattern in South Gojam sub basin was studied by many researchers (Bewket, 2002; Teferi et al., 2013). They reported that the dramatic changes of the natural vegetation cover into the agricultural crop land which has significantly affect the characteristics of the basin.

3.5 CONCLUDING REMARKS

This chapter provides the first sediment balance of the entire Blue Nile catchment at the sub basin scale.

Lack of availability of good quality data reflects on the study accuracy and output reliability. In this study, sediment and flow data were not available in sufficient detail. A field campaign was conducted to collect sediment concentration data from several locations in the whole basin, but only a few locations and limited to the year 2009.

The sediment transport rates were estimated for many locations along Blue Nile River network using rating curves as regression methods. Three approaches were adopted to determine the sediment loads from the rating curves: the normal linear log-log regression, normal log-log regression with correction bias factor and non-linear log-log regression.

Based on the three methods, the long-term annual average sediment load at El Deim was quantified between 130 and 170 million tonne/year. Instead, the long-term average annual sediment loads obtained from the SWAT model at El Deim was found to be 118 million tonne/year. Previous works estimated the sediment at the same location load in 140 million tonne/year. The model therefore seems to underestimate sediment loads by 10 to 30%.

The results of the long-term annual average sediment load showed that Jemma, Didessa and South Gojam sub basins are provided most of the sediment transported by the system.

Land-use change detection based on historical satellite images showed that in all the above mentioned sub basins were subjected to significant land-use change from natural forest to agricultural lands.

Chapter 4
HYDRODYNAMIC CHARACTERISTICS OF THE BLUE NILE RIVER NETWORK[21]

Summary
This chapter describes the water distribution along the Blue Nile River in order to quantify the availability of the water resource at different seasons and flow conditions using a one-dimensional hydrodynamic model covering the entire river system from Lake Tana to Khartoum. The work included an extensive filed measurement campaign along the Blue River and its tributaries, both in Ethiopia and Sudan, to fill in knowledge gaps on hydrodynamic characteristics (cross sections, water levels, discharges) of the river network.

4.1 BACKGROUND

The management of the water resources is increasingly difficult due to the conflicting demands from various stakeholder groups, population growth, rapid urbanization and increasing incidences of natural disasters, especially in arid areas (Kim and Kaluarachchi, 2008). In the Blue Nile River basin, upstream runoff variability is an acute issue to downstream countries, Sudan and Egypt, as they are heavily dependent on the Nile waters. Cooperative management of the Nile waters has become urgent considering increasing demands and climate variations (Conway, 2005; Conway and Hulme, 1996). Moreover, there are new plans for massive water resources developments for hydropower generation in Ethiopia (four large dams on the Blue Nile), as reported by BCEOM et al.(1998). In addition to construction of the Grand Ethiopian Renaissance Dam and the heightening of Roseires Dam, recently the Setit and Bardana dams are under construction across the Atbara River (a Nile tributary) in Sudan.

---

[2] Y.S.A. Ali, A. Crosato, Y.A. Mohamed, S.H. Abdalla, N.G. Wright, J.A. Roelvink (2014) Water resource assessment along Blue Nile River using 1D model, Proceedings of ICE, Water Management Journal. DOI: 10.1680/wama.13.00020.

As a result, the Nile Basin countries are currently negotiating to find a common agreement on how to share the Nile waters (NBI, 2003; NBI, 2004). At present, six countries (Ethiopia, Tanzania, Kenya, Rwanda, Uganda and Burundi) signed the so called cooperative framework agreement (CFA), while Sudan, Egypt, DRC, and South Sudan not yet (BBC, 2010). A tripartite technical committee was formed by the three Eastern Nile countries (Ethiopia, Sudan and Egypt) in the early 2012 to assess the downstream impacts of the Grand Ethiopian Renaissance Dam. In this context, the knowledge of water uses along the Nile River is of basic importance, especially considering that the filling of the vast Grand Ethiopian Renaissance reservoir will occur in the near future. It is also important to consider that water needs vary with the season, but a clear quantitative picture of these variations is still lacking. The Blue Nile contributes 60-65% of Main Nile flow (Sutcliffe and Parks, 1999; UNESCO, 2004; Yates and Strzepek, 1998 b). Therefore every development on this river may affect the Main Nile runoff in a considerable way.

4.2 MATERIALS AND METHODS

4.2.1 Model input data

This study relies on the historical measurements of water levels and flow discharges on a series of gauging stations located along the Blue Nile River. The stations were selected based on two main criteria: their location, ensuring that there is a good coverage of information along the river course and the availability of flow and water level records. The screening of historical data revealed a number of typing mistakes. These were corrected in Excel data files. Missing discharge data for the major tributaries were filled in with a regression equation between neighbouring stations.

The bathymetric survey of 1991 executed by the Ministry of Irrigation and Water Resources and Delft Hydraulics includes 84 cross sections with spacing of 5 to 7 km, covering the reach from Roseires Dam to Khartoum (Delft-Hydraulics, 1992). Other cross sections were measured more recently at scattered locations, such as Singa, Suki, Wad Medani, Khartoum and inside Roseires Reservoir in different years (Abd Alla and Elnoor, 2007; Gismalla, 1993). Different references were used for the topographic data (Tiesler, 2009), for both the vertical and horizontal coordinates. This study uses the UTM WGS-84 coordinate system in horizontal direction for the positions of the cross sections and the Alexandria vertical datum for the vertical levels. To fill in this important knowledge gap, an extensive field survey was conducted as a part of this study in 2009. Flow velocity and bed topography were measured at 26 cross sections along the Blue Nile River and its major tributaries using a current meter and an ecosounder in Ethiopia and an Acoustic Doppler

Current Profiler (ADCP) in Sudan. Figure 4.1 shows the locations of the cross sections surveyed in 2009 and listed in Table 4.1.

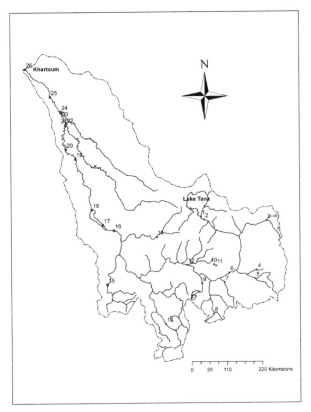

Figure 4.1: Locations of the measured cross sections (denominations in Table 4.1).

Table 4.1: Denomination of measuring station in Figure 4.1.

No	Description	No	Description
1	Downstream Lake Tana	14	Beles River
2	Andassa River	15	Dabus River
3	Jemma River	16	Abay River at Border
4	Welaka River	17	Blue Nile River at Roseires 1
5	Jemma River	18	Blue Nile River at Roseires 2
6	Abay River at Kessie	19	Blue Nile River at Singa
7	Muger River	20	Blue Nile River at Kassab
8	Guder River	21	Dinder River
9	Fincha River	22	Blue Nile River at Nor Al Deen
10	Chemoga River	23	Blue Nile River at Wad Medani
11	Yeda River	24	Rahad River
12	Abay River at Bure	25	Blue Nile River at Hilalia
13	Didessa River	26	Blue Nile River at Khartoum

The river network was divided into three reaches according to the positions of Roseires and Sennar dams. Reach 1 represents the river upstream of Roseires Dam, Reach 2 represents the river between the two dams of Roseires and Sennar and Reach 3 describes the river from Sennar Dam to Khartoum.

As an example, the river cross-sections just downstream of Lake Tana (upstream boundary), at Ethiopian-Sudanese border (middle) and at Khartoum (downstream boundary) are depicted in Figure 4.2 together with an image of the river at the same location.

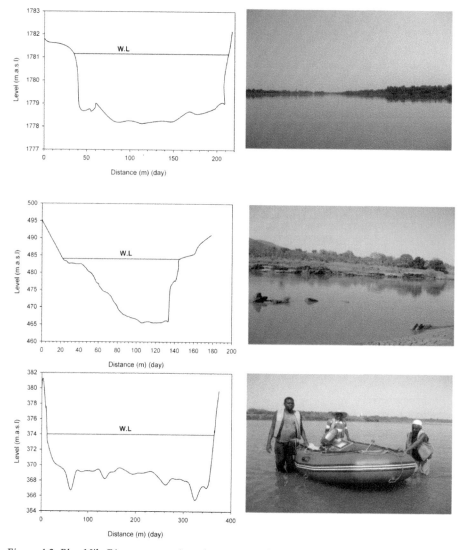

Figure 4.2: Blue Nile River cross sections downstream of Lake Tana, at the Ethiopian-Sudanese border and at Khartoum. Note that the graphs have different horizontal and vertical scales.

4.2.2   Model development

The flow in the river network was modelled using the SOBEK RURAL package (www.deltares.nl). This is a one-dimensional open-channel dynamic numerical modelling system for water flow in open channels, which has been successfully applied on river systems all over the world (Prinsen and Becker, 2011). The model is based on the Saint-Venant equations for unsteady flow as stated below:

$$\frac{\partial A}{\partial t} + \frac{\partial Q}{\partial x} = 0 \qquad\qquad \text{(continuity equation)} \qquad\qquad 4.1$$

$$\frac{\partial Q}{\partial t} + \frac{\partial}{\partial x}\left[\alpha \frac{Q^2}{A}\right] + gA\frac{\partial h}{\partial x} + \frac{gQ|Q|}{C^2RA} = 0 \qquad \text{(momentum equation)} \qquad 4.2$$

Where: A is the wet cross-sectional surface (m²); Q is the discharge (m³/s); t is time (s); x is the longitudinal distance (m); α is the Boussinesq coefficient (-); g is the acceleration due to gravity (m/s²); h is the water depth (m); C is Chézy coefficient (m$^{1/2}$/s) and R is the hydraulic radius (m).

The computation of the water levels and discharges in the SOBEK-flow-network is performed with the Delft-scheme. This scheme solves the Saint-Venant equations (continuity and momentum equations) by means of a staggered grid in which the water levels are defined at the connection nodes and calculation points, while the discharges are defined at the intermediate reaches or reach segments. The software allows for the inclusion of several types of hydraulic structures such as weirs, sluice gates, pumps and locks as well as their operation rules.
The domain of the developed model covers the Blue Nile River from the outlet of Lake Tana (upstream boundary) to Khartoum (downstream boundary). The length of the main stream of the river is about 1,600 km, about 950 km in Ethiopia and 650 km in Sudan.
A large part of cross-sectional data from the Sudanese part of the river dates from 1990, but unfortunately the only cross-sections available for the Ethiopian part of the river are those measured in 2009 in the framework of this project. The total number of cross-sections available for the entire Blue Nile River network is 168.
To allow sufficient accuracy, the space step of the longitudinal (1D) grid was set equal to 3,000 m. The value of the time-step was derived imposing the Courant condition:

$$c\frac{\Delta t}{\Delta x} \leq 10 \qquad\qquad\qquad\qquad 4.3$$

Where: c is the celerity of the flood wave (m/s).

According to the available data, at El Deim station:
- Average flow width: 500 m
- Maximum daily change in discharge in the rising limb: 1500 m³/s
- Water level change at the maximum daily discharge change: 0.75 m
- Celerity of the flood waves: 4 m/s.

Based on Equation 4.3, the time step must satisfy the following condition: $\Delta t < 7,500$ s (125 minutes), so in the model, the time step was set to one hour.

The daily outflow discharge from Lake Tana at Bahridar define the upstream boundary conditions of the hydrodynamic model, whereas the daily discharge time series from the major tributaries are used as lateral flows entering the system. Water extraction for the irrigation schemes described in chapter 2 are used as lateral flows leaving the system. The time series water levels at Khartoum station define the downstream boundary conditions.

The model includes the dams of Sennar and Roseires with their operation rules. The model schematization, including the dams and of the hydrological stations is depicted in Figure 4.3.

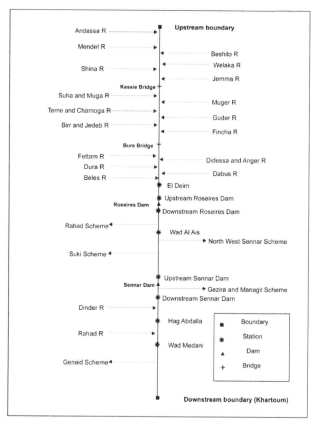

Figure 4.3: Model schematization including lateral flow entering the system, water withdrawal for irrigation schemes and measuring stations.

4.2.3   Model calibration and validation

The uncertainties related to the values of the variables describing the hydrodynamic behaviour of the Blue Nile network and the simplifications in the numerical description of the processes involved create the necessity to calibrate the model. The idea underlying the calibration processes is to reduce the uncertainties by matching model results with the available measurements. The model should be able to capture the main discharge characteristics through the river network and to reproduce the relative water level fluctuations along the river network.

Model calibration and validation were carried out by comparing computed to measured water levels at El Deim, just upstream of Roseires Dam, Wad Al Ais, just upstream of Sennar Dam (inside the reservoir), Hag Abdalla and Wad Medani (Table 4.2) and by comparing computed to measured discharges at Kessie Bridge, El Deim, just downstream of Roseires Dam and just downstream of Sennar Dam station (Table 4.3).

Table 4.2: Water level calibration and validation chaining.

Branch number	No	Station Name	Distance upstream Khartoum (km)
Reach 1	a	El Deim	740
	b	Upstream Roseires Dam	630
Reach 2	c	Wad Al Ais	436.5
	d	Upstream Sennar Dam	351
Reach 3	e	Hag Abdalla	282.5
	f	Wad Medani	200.5

Table 4.3: Discharge calibration and validation chaining.

Branch number	No	Station Name	Distance upstream Khartoum (km)
Reach 1	a	Kessie Bridge	1340
	b	El Deim	740
Reach 2	c	Downstream Roseires Dam	629
Reach 3	d	Downstream Sennar Dam	345

The primary parameter required for the calibration of SOBEK RURAL is the bed roughness (McGahey et al., 2012). The Manning-Stickler coefficient was chosen for the calibration instead of the Chezy coefficient. The model can freely calculate the Chezy coefficient considering the local hydraulic radius taken from the last iteration loop. The Strickler formula is one of the methods to define the bed roughness. The actual value of Chezy coefficient is computed using:

$$C = k_s R^{1/6} \qquad\qquad 4.4$$

Where: k_s is the Manning-Strickler coefficient in $m^{1/3}/s$ and R is the hydraulic radius in m.

The calibration was performed by adjusting the values of k_s in order to get a good reproduction of the observed water level and discharges at a number of gauging stations.

The model was calibrated for the period 1990-1993. Several runs were carried out with different Manning-Strickler coefficients, ranging between 30 $m^{1/3}/s$ and 70 $m^{1/3}/s$ with the aim to optimize the reproduction of the observed water levels and discharges at a number of gauging stations. The adjusted values of Manning-Strickler coefficients were found to be 60, 55, 50 and 40 $m^{1/3}/s$ in the reach from Lake Tana to El Deim, from El Deim to Roseires dam, between Roseires Dam and Sennar Dam and in the reach between Sennar Dam and Khartoum, respectively.

The results of model calibration in terms of discharges and water levels are shown in Figures 4.4 and 4.5, respectively. The comparison between computed and measured water levels shows a general tendency to slightly underestimate high-flow levels and slightly overestimate low-flow levels inside the reservoirs of Roseires and Sennar, whereas the model performs well for the other stations along the river. The comparison between modelled and measured discharges shows a good agreement between model and discharge measurements.

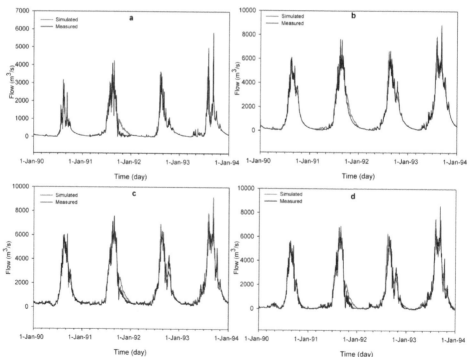

Figure 4.4: Comparison between observed and simulated discharges at (a) Kessie Bridge, (b) El Deim, and from (c) Roseires and (d) Sennar Dam. Period January 1990 - January 1994.

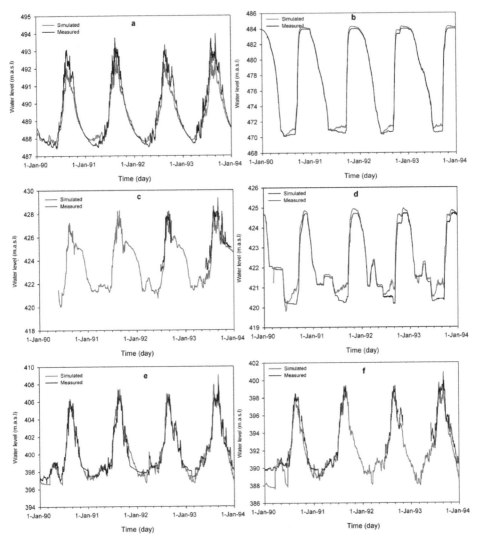

Figure 4.5: Comparison between observed and simulated water levels at (a) El Deim, (b) upstream of Roseires Dam, (c) Wad Al Ais, (d) upstream of Sennar Dam, (e) Hag Abdalla and (f) Medani. Period January 1990 - January 1994.

Because the calibration process involves some adjustments of parameter values that are optimized to fit a certain data set, good model calibration cannot automatically ensure that the model performs equally well also for other periods and circumstances. Therefore, model validations on independent data are required. The model was therefore validated comparing its results to the measured data for the three years 1994, 1995 and 1996. The discharge and water level results are shown in Figures 4.6 and 4.7 respectively. Again the model shows a slight under prediction of peak-flow levels and a slight over prediction of low-flow levels inside the reservoirs of Roseires and Sennar, whereas the model performs well for the other stations along the river. The model predicted well the discharges during low flows, but overestimated the highest discharges at all stations, as shown in Figure 4.7.

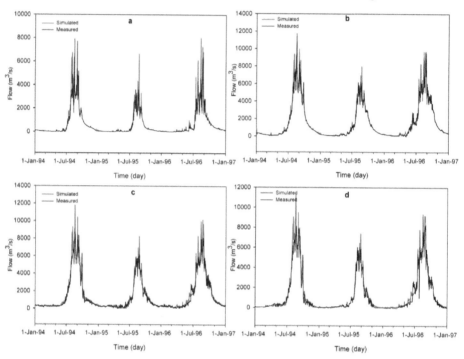

Figure 4.6: Comparison between observed and simulated discharges at (a) Kessie Bridge, (b) El Deim, and from (c) Roseires and (d) Sennar Dam. Period January 1994 - January 1997.

There are several statistical measures that can be used to evaluate the performance of simulation models, among which the assessment of the correlation coefficient (R^2), the Nash-Sutcliffe model efficiency (NSE) (Nash and Sutcliffe, 1970) and the root mean square error (RMSE). The correlation coefficient (R^2) reflects the linear relationship between observed and simulated data and is thus insensitive to either an additive or a multiplicative factor. Some authors recommend this parameter when large-scale models with gridded fields are involved. R^2 ranges from 0 to 1, with higher values indicating less error variance, and typically values greater than 0.5 are considered acceptable (Santhi et al., 2001; Singh et al., 2004; van Liew et al., 2003).

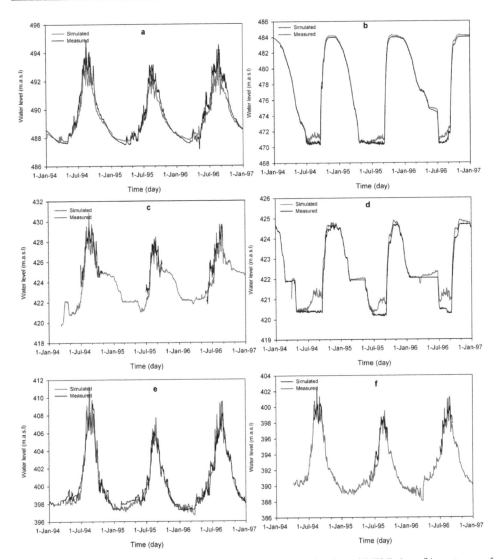

Figure 4.7: Comparison between observed and simulated water levels at (a) El Deim, (b) upstream of Roseires Dam, (c) Wad Al Ais, (d) upstream of Sennar Dam, (e) Hag Abdalla and (f) Medani. Period January 1994 - January 1997.

Correlation coefficient, R^2 is given by the equation

$$R^2 = \frac{\sum_{i=1}^{n}(x_i-\bar{x})^2(y_i-\bar{y})^2}{\sum_{i=1}^{n}(x_i-\bar{x})^2 \sum_{i=1}^{n}(y_i-\bar{y})^2} \qquad\qquad 4.5$$

where x_i is the simulated value and y_i is the actual value.

The Nash-Sutcliffe model efficiency (NSE) indicates how well the plot of observed versus simulated data fits the 1:1 line; it is computed according to the following equation:

$$NSE = 1 - \frac{\sum_{j=1}^{n}(y_j - x_j)^2}{\sum_{i=1}^{n}(y_i - \bar{y})^2} \qquad\qquad 4.6$$

The root mean square error (RMSE) is a commonly used error index (Chu and Shirmohammadi, 2004; Singh et al., 2004; Vazquez-Amábile and Engel, 2005). RMSE gives the standard deviation of the model prediction error. Values close to zero indicate a perfect fit. Its value should be compared to the standard deviation of the measured data. Values less than half of the standard deviation of the measured data are considered acceptable (Singh et al. (2004)). RMSE is given by the equation:

$$RMSE = \sqrt{\frac{(x_1 - y_1)^2 + (x_2 - y_2)^2 + \cdots + (x_n - y_n)^2}{n}} \qquad\qquad 4.7$$

The RMSE index, NSE and correlation coefficients were computed to quantify the model performance. Tables 4.4 and Table 4.5 showed the values of these indexes for the simulations of the water levels and the discharges in the periods 1990-1993 (calibration) and 1994-1996 (validation), respectively.

Model calibration resulted in correlation coefficients between 0.947 and 0.999 and NSE values between 0.910 and 0.999 for the water levels. For the discharges, the correlation coefficients fell between 0.95 and 0.972 and the NSE values between 0.044 and 0.970.

Model validation resulted in correlation coefficients between 0.93 and 0.998 and in NSE values falling between 0.926 and 0.995 for the water levels. For the discharges the correlation coefficients fell between 0.894 and 0.964 and the NSE values between 0.880-0.959. The results showed that RMSE values are always less than 50% of the standard deviation of the measured data (Tables 4.4 and 4.5). The model performance can therefore be considered good.

Table 4.4:.Water level, calibration and validation correlation coefficient and Root Mean Squared Error.

Name	Calibration				Validation			
	R²	NSE	RMSE	SD	R²	NSE	RMSE	SD
El Deim	0.977	0.91	0.26	1.73	0.978	0.926	0.27	1.81
Upstream Roseires	0.999	0.995	0.2	5.55	0.998	0.995	0.22	5.36
Wad Al Ais	0.999	0.999	0.39	2.13	0.955	0.947	0.47	2.19
Upstream Sennar	0.975	0.955	0.25	1.59	0.974	0.947	0.26	1.6
Hag Abdalla	0.953	0.943	0.59	2.7	0.951	0.943	0.67	3.02
Wad Medani	0.947	0.944	0.69	3.02	0.93	0.927	0.8	3.01

Table 4.5: Discharges, calibration and validation correlation coefficient and Root Mean Squared Error.

Name	Calibration				Validation			
	R²	NSE	RMSE	SD	R²	NSE	RMSE	SD
Kessie Bridge	0.96	0.958	155.82	777.05	0.894	0.88	341.29	1049.17
El Deim	0.972	0.97	295.97	1783.05	0.964	0.959	385.59	2018.27
Downstream Roseires	0.959	0.958	363.92	1790.43	0.943	0.936	470.35	1965.61
Downstream Sennar	0.95	0.944	387.45	1730.54	0.934	0.932	524.08	2043.13

4.3   ASSESSMENT OF CURRENT WATER DISTRIBUTION

Since most data used for model set up, calibration and validation are from the 1990's, the model strictly represents the Blue Nile river behaviour 15-20 years ago. To check its performance on more recent years, the model was applied to study water levels and discharge distribution in 2008, 2009 and 2010. The results are shown in Figure 4.8 and Figure 4.9. It is possible to observe that the agreement between simulated and measured water levels and discharges is still good. Similarly to the calibration and validation runs, the model tends to over predict low-flow water levels inside the reservoirs of Roseires and Sennar, but now also in the reach between the two dams, as well as from Sennar Dam to Khartoum, and to over predict peak discharges. This model behaviour can be an indication of unknown water uses or water losses during the dry season, but this can be established only by analysing the model performance considering its simplifications and shortcomings.

As described in chapter 2, several irrigation schemes in Sudan withdraw water from the Blue Nile River, especially in the reach between the dams of Roseires and Sennar and from Sennar to Khartoum. The model includes the water extractions by the irrigation schemes, but does not include the additional small (private) schemes, since it is difficult to assess their amount of abstraction. For this, pumping rates and pumping hours should be used as inputs to precisely calculate their consumption, but no one knows these variables and there are thousands of small pumping schemes.

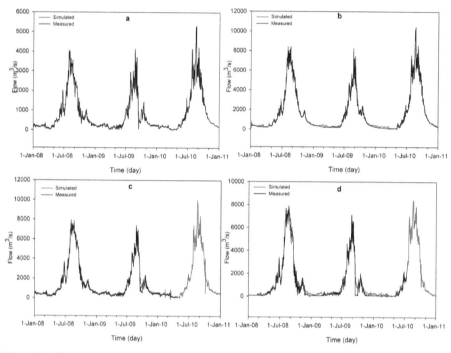

Figure 4.8: Comparison between observed and simulated discharges at (a) Kessie Bridge, (b) El Deim, and from (c) Roseires and (d) Sennar Dam. Period January 2008 - January 2011.

Over prediction of low-flow water levels can be caused by not taking into account these water extractions, but it can also be due to systematic overestimation of the channel bed roughness during low flows. In the model, the roughness varies as a function of water depth without taking into account any additional flow resistance due to dune formation. For the particular case of the sand-bed Blue Nile, we can expect that dunes form at low flows but not at high flows when the concentration of suspended solids is very high. This means that most probably the Blue Nile river bed is much rougher during low flows than during high flows. The effects of using one single value of Manning Strickler coefficient means that the channel roughness during low flows is most probably underestimated rather than overestimated by the model. Based on this, the discrepancy between measured and computed water levels during the low flow season appears to be related to losses of water, which can be caused by evaporation (not included in the model), ground water recharge (not included in the model) and water withdrawal (partly included in the model), rather than to bed roughness overestimation.

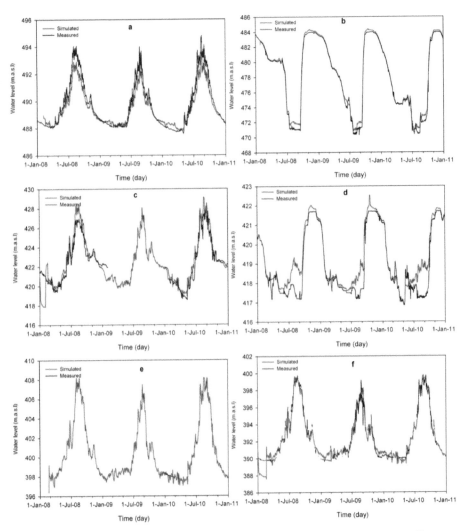

Figure 4.9: Comparison between observed and simulated water levels at (a) El Deim, (b) upstream of Roseires Dam, (c) Wad Alais, (d) upstream of Sennar Dam, (e) Hag Abdalla and (f) Medani. Period January 2008 - January 2011.

It is difficult to thoroughly quantify the losses due to ground water recharge, but it can be reasonably assumed that most of these losses are due to the water abstractions that have not been quantified. The model was finally used to quantify the total water losses in the region between Roseires and Khartoum, which include water extraction and ground water recharge. These were found to be about 15% of the known extracted volumes.

4.4 CONCLUDING REMARKS

This study deals with the construction of a 1D hydrodynamic model of the entire Blue Nile River system to be used to quantify the availability of the water resource throughout the year.

The model was built on cross-sectional data from the 1990s for the Sudanese part of the river and from 2009 for the Ethiopian part. It was calibrated on the period 1990 to1993 and validated on the period 1994 to 1996. The simulations showed good correlation coefficients between computed and measured water levels and discharges for both the calibration and validation runs.

These results give confidence in the model, particularly considering the need to estimate discharges from measured water levels at locations where discharge measurements are scarce or absent. However, since most data used for model set up, calibration and validation are from the 1990's, the model strictly represents the Blue Nile river behaviour 15-20 years ago. To check its performance on recent years, the model was then applied to study water levels and water distribution in the recent years 2008, 2009 and 2010.

The comparison between computed and measured water levels and discharges shows that it performs well also for the present situation. For this, the morphological changes occurred in the last 20 years do not seem to have affected the hydrodynamic behaviour of the Blue Nile River system in a big way.

The Blue Nile River Basin is currently subject to many challenges, such as growing population, which will increase the water demand, and dam construction, which may affect the river regime in particular during low flows. Critical moments related to the availability of the water resource can be expected in the future during the filling phase of the newly constructed reservoirs.

Chapter 5
SEDIMENT FINGERPRINTING IN ROSEIRES RESERVOIR[3]

Summary
This chapter aims to link the sediment deposited inside Roseires Reservoir (sink) with the sediment from the upper basin. First, sedimentation processes inside Roseires Reservoir were simulated to identify promising coring areas for the study of soil stratification in the reservoir. Two field campaigns were followed, one inside Roseires Reservoir and another one in the upper basin to study the characteristics of the deposited sediment in the reservoir and eroded sediment respectively. Samples were analyzed using X-Ray Diffraction (XRD) to identify the mineral content, and then cluster analysis was applied to identify groupings of objects that share a "similarity", which can be quantified in terms of any measurable parameter. Cluster analyses were performed on the X-Ray Diffraction (XRD) data in Minitab software package.

5.1 BACKGROUND

Weathering is the physical, chemical, and biochemical breakdown of Earth materials at the interface between the lithosphere and the atmosphere (Allen, 1997). It is the starting point of sediment creation by producing a soil: a layer of loose, unconsolidated sediment of variable thickness over the land surface that hosts plant growth in all but the driest, coldest, and most saline areas. Weathering determines in large part the initial mineralogy, size, and shape of the sediment grains eroded and transported out of a source area (Allen, 1997; Hefferan and O'Brien, 2010).

Many methods are used for semi qualitative identification of minerals in geological samples by finger-printing approach, one of them is XRD (Al-Jaroudi et al., 2007; Xie et al., 2013), this method is proven effective in the quantification of mineralogical data (Ottner et al., 2000; Ruan and Ward, 2002). The intensities from an individual mineral are used to quantify mineral content in the sample (Cullity and Stock, 1956).

---

[3] OMER A.Y.A., ALI Y.S.A., CROSATO A., PARON P. ROELVINK J.A. & DASTGHEIB A., 2014. Modelling of sedimentation processes inside Roseires Reservoir (Sudan). Earth Surf. Dynam. Discuss., 2, 153-179, www.earth-surf-dynam-discuss.net/2/153/2014/ doi:10.5194/esurfd-2-153-2014.

Measurement of peak intensities do, therefore provide information regarding the relative amount of the corresponding mineral phase in each sample.

Sediment fingerprinting links the mineralogical or geochemical properties of the sediment to its source material. If source materials can be distinguished by their geochemical properties, the likely source of the sediment can be established by comparing the properties of the sediment with source materials (Walling et al., 2003). The need to discriminate several potential sediment source areas means that a single fingerprint property is generally unlikely to provide a reliable source fingerprint. Therefore, most recent source fingerprinting studies have used composite fingerprint, comprising a range of different diagnostic properties and mixing models to quantify the relative contributions of sediment from different sources (Collins and Walling, 2002; Collins et al., 2010).

Cluster analysis is a powerful tools for classifying and sorting data to establish relationship within such data (Sneath and Sokal, 1973; Yang and Simaes, 2000). Cluster analysis is also called segmentation analysis or taxonomy analysis (Aldenderfer and Blashfield, 1984; Everitt et al., 2001). The method creates groupings of objects that share a "similarity", which can be quantified in terms of any measurable parameter. Many different fields of study, such as engineering, zoology, medicine, linguistics, anthropology, psychology, marketing, and indeed geology have contributed to the development of clustering techniques and the application of such techniques (Cortés et al., 2007; de Meijer et al., 2001; Mamuse et al., 2009).

Two cluster analysis approach can be performed: (1) hierarchical clustering (Johnson, 1967; Kaufman and Rousseeuw, 2009), in which the data are grouped using an iterative algorithm into clusters (2) K-means clustering (Army, 1993; Kanungo et al., 2002; Wagstaff et al., 2001), in which the number of clusters is defined a priori and all the data points are distributed into the clusters on the basis of some particular characteristic or metric. In this study the hierarchical clustering was used to create cluster tree called dendrogram that allows deciding the level or scale of clustering that is most appropriate for an application. There are several methods to perform hierarchical clustering such as:

1. Single linkage method, is based on a hierarchy built using the smallest distance between one of the individuals within one cluster to one of the individuals in adjacent clusters. This methodology is useful to identify irregular cluster shapes but is limited by poor performance in statistical tests and by the fact that the graphic representation of the hierarchical tree is difficult to interpret to gain direct quantitative information about the complete cluster size and shape.

2. Complete linkage method is used to build up the cluster hierarchy and consists of finding the proxy of a cluster, based on finding the data point within the cluster which shows the largest distance between itself and its nearest surrounding cluster. This technique tends to identify clusters that are equi-dimensional. Also in this technique, the maximum distance between one cluster and another corresponds to the maximum size of the new cluster, thus representing the maximum cluster size at each partition.

3. Centroid linkage method uses the distance between the centroids of two groups, usually calculated using the arithmetic mean. The principal caveat with this method is that it can produce a cluster tree that is not monotonic. This occurs when the distance from the union of two clusters (a and b) to a third cluster is less than the distance from either a or b to that third cluster.

4. Average linkage uses the average distance between all pairs of objects in clusters. The main advantage of this method is that being an analogue to the centroid method, in the sense that it is also an estimation of the central tendency of the data and it produces monotonic hierarchical trees. This method is often used with a Euclidean metric as a measure of similarity. This method was used in this study.

Three main geological units have been distinguished in the upper Blue Nile in Ethiopia: (i) Precambrian metamorphic rocks represent the oldest complex. They form the crystalline basement underlying the volcanics, and are exposed over about 32 % of the basin in the western lowlands; (ii) Mesozoic sedimentary formations represent about 10-11 % of the basin, being mainly visible in the deep valleys of major southern tributaries of the Upper Blue Nile and (iii) thick Tertiary and Quaternary volcanic formations cover 55-56 % of the basin, in the north, centre and east of the basin (MWR, 1999; Wolela, 2007; Wolela, 2008). The geology map for the entire Blue Nile River Basin is depicted in Figure 5.1.

The main part of the Sudan Lowlands (Figure 5.1) is underlain by deep unconsolidated colluvial sediments of tertiary and Quaternary age. To the north are older Basement Complex rocks and the Nubian Sandstones. The Nubian Sandstones are located in the northwest corner and overly uncomfortably the Basement Complex rocks and comprise mainly sandstones, siltstones and conglomerates(ENTRO, 2007).

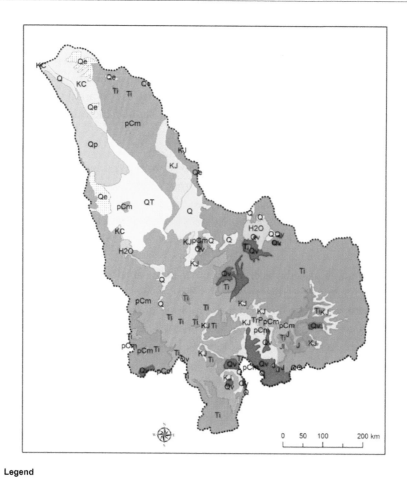

Legend

——— Stream

Blue Nile Geology

Geological formations

▢ Water bodies

▢ Q (alluvial, lacustrine and swampydeposits)

▒ Qe (Holocene sands and loose sediments)

▢ Qp (unconsolidated clays, silts, sands and gravel)

▢ QT (unconsolidated sands with some gravels, clays, and shales)

▢ Qv (effusive rocks: rhyolitic and basaltic lavas, ignimbrites, lacustrine and swamp deposits))

▢ KJ (sandstone, evaporites, limestones, marls)

▢ J (sandstone, evaporites, limestones, marls)

▢ Jl (sandstone, evaporites, limestones, marls)

▢ KC (Nubian sandstone formation: continental clastic sediments including sandstones, siltstones, mudstones, and conglomerates)

▢ Ti (Highland volcanic rocks)

▢ TrP (basement rocks - metamorphic)

▢ pCm (Undifferentiated basement rocks - metamorphic)

▢ Blue Nile Basin

Figure 5.1: The geology map for the entire Blue Nile River Basin.

5.2   SEDIMENT SORTING INSIDE ROSEIRES DAM

Sedimentation processes inside Roseires Reservoir were analyzed during the period from 1985 to 2007 using a physics-based quasi 3D morphodynamic model (Delft3D)(Omer, 2011). The setup of the model was constructed by developing a 2D depth-averaged hydrodynamic model; followed by a morphodynamic model considering two types of sediments: fines (supply-limited) and coarse (capacity limited), capable of storing the history of bed level changes as well as bed composition (bed layer model).

The hydrodynamic model that reproduces flow velocities and water levels of the system composed by Roseires Reservoir and two short reaches of the Blue Nile River, upstream and downstream of the reservoir was calibrated and validated on daily time series of inflow and outflow discharges and water levels near the dam relative to the years 2009 and 2010.

The morphodynamic model was based on the 2D hydrodynamic model, using on-line method with morphological factor (Roelvink, 2006). It was calibrated and validated on the observed bed level changes during two distinct periods: 1985-1992 and 1992-2007.

5.2.1   Model description

Lesser et al. (2004) extensively describe the Delft3D model which is applied in the current study (Omer, 2011). The model solves the shallow water equations and calculates sediment transport based on the generated flow field, using different type of sediment transport relations. In this study a 2D depth-averaged schematization was used with appropriate quasi-3D formulations for the spiral flow in bends. Bed elevation is updated every time step based on the sediment transport field. Since morphodynamic developments are typically much slower than the hydrodynamic processes, the calculated bed level change is multiplied by a morphological factor (MF) every time step to enhance morphodynamic development (Roelvink, 2006).

The model includes a slightly adapted version of the Hirano (1971) bed layer model, as it is described in Blom (2008). Similar to the grid of the hydrodynamic numerical model, the bed is subdivided into separate cells. A bed cell consists of different numerical layers. These layers allow for different specifications of bed composition and sediment characteristics and record the changes of bed composition in the vertical during the morphodynamic simulations.

The hydrodynamic part of the model is based on the 3-D Reynolds-averaged Navier–Stokes (RANS) equations for incompressible fluid and water (Boussinesq approximation). We used a 2-D depth-averaged version of the model with an appropriate parameterization of two relevant 3-D effects of the spiral motion that arises in curved flow (Blanckaert et al., 2003). In addition, Delft3D model has capability to record bed changes in z-direction (Quasi-3D). First, the model corrects the direction of sediment transport through a modification in the direction of the bed shear stress, which would otherwise coincide with the direction of the depth-averaged flow velocity vector. Second, the model includes the transverse redistribution of main flow velocity due to secondary-flow convection, through a correction in the bed friction term. The closure scheme for turbulence is a k-ε model, in which k is the turbulent kinetic energy and ε is the turbulent dissipation.

The evolution of bed topography is computed from a sediment mass balance, a sediment transport formula for capacity-limited sediment transport, as well as an advection-diffusion formulation of suspended solids concentrations for fine supply-limited sediment transport, coupled to two formulas describing the entrainment and deposition processes.

The equations are formulated in orthogonal curvilinear co-ordinates. The set of partial differential equations in combination with the set of initial and boundary conditions is solved on a finite-difference grid.

A number of capacity-limited transport formulas are available, such as Meyer-Peter and Muller's (1947), Engelund and Hansen's (1967) and van Rijn's (1984). The model accounts for the effects of gravity along longitudinal and transverse bed slopes on bed load direction (Bagnold, 1966; Ikeda, 1982).

The concentrations of fine sediment in suspension are calculated by solving the following 3D advection-diffusion equation (mass balance):

$$\frac{\partial c}{\partial t} + \frac{\partial uc}{\partial x} + \frac{\partial vc}{\partial y} + \frac{\partial (w-w_s)c}{\partial z} = \frac{\partial}{\partial x}\left(\varepsilon_{s,x}\frac{\partial c}{\partial x}\right) + \frac{\partial}{\partial y}\left(\varepsilon_{s,y}\frac{\partial c}{\partial y}\right) + \frac{\partial}{\partial z}\left(\varepsilon_{s,z}\frac{\partial c}{\partial z}\right) \qquad 5.1$$

Where: c is the mass concentration of the fine sediment fraction (kg/m³) and u, v and w are the flow velocity components (m/s). The velocity and eddy diffusivity ($\varepsilon_{s,x,y,z}$) are gained from the hydrodynamic computations (continuity and momentum conservation equation for water). In computing the settling velocity of the sediment particles (w_s), the hindered effect is taken into account. The adjusted settling velocity is given by the following formula:

$$w_s = \left[1 - \frac{c_s}{c_{soil}}\right]^5 w_{s,0} \qquad 5.2$$

in which: c_{soil} is the reference density for hindering settling (kg/m³) and c_s is the total mass concentration of the fine sediment fractions (kg/m³).

The following general formula (Ariathurai and Arulanandan, 1978; Partheniades, 1964) describes the entrainment of fine sediment from the bed:

$$E = M\left[\frac{\tau-\tau_c}{\tau_c}\right] \qquad 5.3$$

Where: E is the erosion flux (kg/m²/s) and M is a coefficient quantifying the quantity of entrained sediment (kg/m²/s),

The following formula describes the deposition rate:

$$D = C_a.w_s \hspace{10cm} 5.4$$

In which: D is the deposition rate ($kg/m^2/s$) and C_a is the sediment concentration
near the bed.

In this study the model is simplified into 2D horizontal (depth-averaged)
formulations. The implications for fine sediment concentrations are described in
Montes et al. (2010).

5.2.2   Hydrodynamic model setup, calibration and validation

A 2D depth-averaged model was built to cover the Reservoir area and extends to El
Deim station (about 110 km from the dam and 30 km upstream the end of the
reservoir) (Omer, 2011). The building of the computational grid was executed
considering basic principles of discretisation such as the schematization of hydraulic
and morphological processes and computational time of morphological simulation.
The basic assumption of orthogonality and smooth changes of cell width and length
were considered, in order to avoid non-accuracies of the numerical computations
during dynamical procedures. The reservoir shape is rather complex, consequently,
the computational grid size varied between 25 to 280 m from El Deim station
upstream to the dam downstream respectively (Figure 5.2).

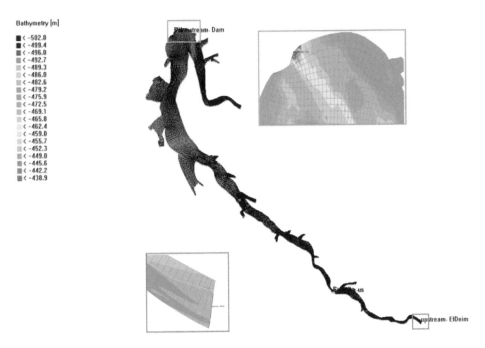

*Figure 5.2: Upstream and downstream boundaries, computational grid and bed elevations in 2009, in
m above sea level (Irrigation datum).*

The hydrodynamic model was calibrated and validated based on the 2009 and 2010 measurement. The calibration and validation process was carried out by comparing the simulated and measured water level at Famaka Station (about 80 km upstream of the dam). Figure 5.3 shows satisfactory results of calibration and validation of the hydrodynamic model.

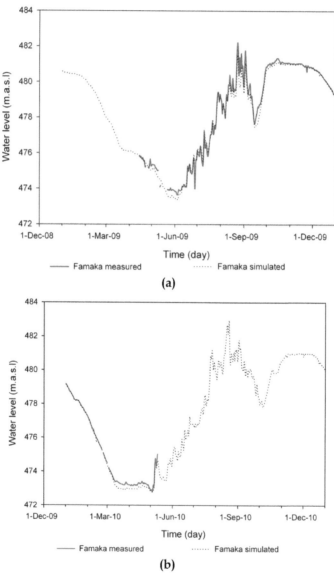

Figure 5.3: Results of hydrodynamic model calibration (a) and validation (b) at Famaka station.

5.2.3   Morphodynamic model setup, calibration and validation

The calibrated hydrodynamic model was used to set up the morphodynamic model. The morphodynamic model was set up for a period of seven years (1985-1992). The upstream inflow and outflow of suspended sediment concentrations were represented by the measured suspended sediment during high flows at El Deim and downstream dam as presented in section 2.4 in chapter 2. The measured suspended sediment D_{50} is around 18.5µm, which gives with a fall velocity of 0.17 mm/s. The critical shear stress for erosion, the erosion rate parameter and the dry density of the deposited sediment were calibrated. Since the D_{50} of the bed material is strongly variable (between 1200µm and 140µm), this parameter was also used as calibration parameter. The simulation of a month morphodynamically is represented by one day hydrodynamically, by simulating a year with twelve days and using a morphological factor of 30.167. This gives a possibility to simulate 84 days and represent seven years of morphology changes at the reach. The model was set up and calibrated on a period of seven years from 1985 to 1992. For the morphodynamics, monthly-averaged inflow and outflow discharge and water levels were used, instead of the daily values of the hydrodynamic model.

The model was calibrated by comparing computed bed topographies with the measured ones at the scale of single cross-sections and at the scale of the entire reservoir (total sedimentation rates). The calibration parameters were the sediment characteristics. Calibration and validation were based on the period 1985-1992 and 1992-2007 respectively. The transport formula that gave the best results for the sand component was Van Rijn's (1984). Calibration resulted in a D_{50} of 700 µm for sand, in a dry density of the sand deposit of 2000 kg/m³ (considering the high compaction rates due to the strong variability of sediment size). The fall velocity of silt was reduced to 0.005 mm/s, whereas the dry density of silt deposits resulted 1200 kg/m³. The choice of this value was based on previous work (Middelkoop, 2002; Middelkoop and Asselman, 1998). The critical shear stress for erosion of fine sediment from the reservoir bed resulted 1 N/m², the erosion parameter rate resulted 2 mg/m²/s (in Equation 5.3).

The results of calibration were illustrated in Figure 5.4. Figure 5.4 (a) showed the measured cumulative erosion and deposition occurred the calibration period (difference between the topography of 1985 and 1992). Figure 5.4 (b) showed the simulated cumulative erosion and deposition for the same period. In the figures the ellipses show the areas for which the bed topography of 1985 and 1992 was unknown and put equal to the bed topography 2009.

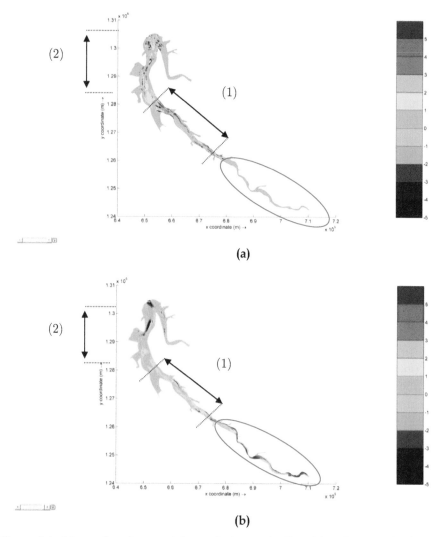

Figure 5.4: Measured and computed cumulative erosion/deposition (in meters): (a). measured topography 1992 – topography 1985 and (b) simulated topography 1992 – topography 1985.

Figure 5.5 showed measured and simulated cross-sections 18 and 19B (10.8 km and 15.4 km upstream of the dam). The model did not provide accurate results at this level of detail. Computed section 18 (Figure 5.5 above) showed that the model fails to simulate main channel shift (compare measured and simulated 1992 topography). The same applied to Section 19B (Figure 5.5 below).

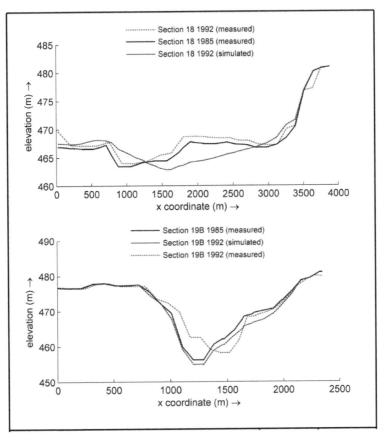

Figure 5.5: Cross-sections 18 and 19B seen from downstream. Measured 1985 and 1992 bed elevations and simulated 1992 bed elevation.

The model was validated on a period fifteen years, from the end of 1992 to the end of 2007. The results were compared to the measured bed levels in 2007. The boundary conditions were the time series of monthly inflow and outflow discharges and averaged water levels inside the reservoir (pool water levels) of these years. The bed topography of 1992 described the initial bed elevation. The upstream inflow and outflow of suspended sediment concentrations were represented by the measured suspended sediment during high flows at El Deim and downstream dam as presented in section 2.4 in chapter 2. Computed cumulative deposition and erosion (in m) for the period 1992 to 2007 are shown in Figure 5.6 (a), and the difference between the measured bed topography of 2007 and 1992 data was illustrated in Figure 5.6 (b).

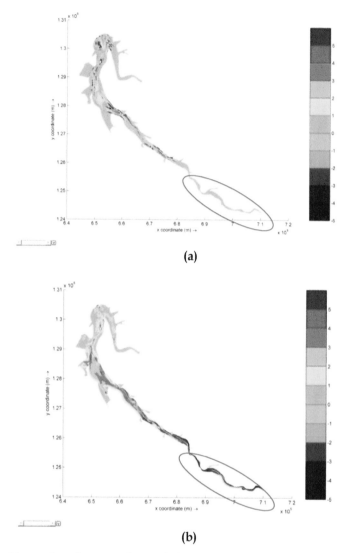

Figure 5.6: Measured and computed cumulative erosion/deposition (in meters): (a). measured topography 2007 – measured topography 1992; (b) simulated topography 2007 – measured topography 1992. The area enclosed by the ellipse missed topographic data of 1992 and therefore cannot be used for comparison.

The comparison between the measured and simulated sections 18 and 19B is shown in Figure 5.7. The simulated section 18 in 2007 showed deposition of 2-2.5 m with respect to section 18 of 1992 which is larger than the measured one. Again, the model is not capable of correctly reproduce main channel shift inside the reservoir at section 19B.

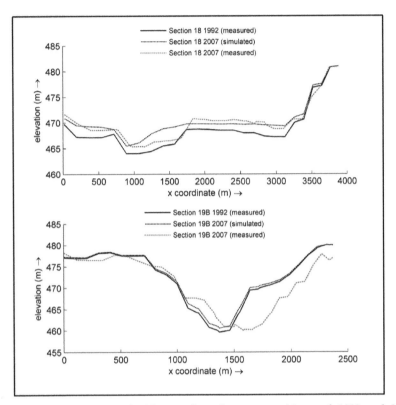

Figure 5.7: Cross-sections 18 and 19B seen from downstream. Measured 1992 and 2007 bed elevations and simulated 2007 bed elevation.

5.2.4   Identification of promising coring locations

Coring locations must fulfil three conditions: 1) absence of net bed erosion; 2) absence of bar movement, destroying soil stratification; 3) recognizable soil stratification, ideally by yearly alternation of sand and silt layers. Stratification should at least allow for the recognition of specific extreme wet or extreme dry years, as for instance 1988.

The model shows variable sedimentation rates during the 22 years between 1985 and 2007. In particular, in 1988 the reservoir gained more storage capacity. This can be referred to the extremely high flood of 1988, which caused net erosion, especially in the area just upstream of the dam.

Reservoir soil stratification should be studied in the areas that were always subjected to deposition. In Roseires Reservoir, a main natural channel having a sinuous shape meanders through the deposited sediment. This channel was found to shift laterally at certain locations. The low model resolution, caused by the relatively large model grid size, made it difficult to correctly detect the channel movement inside the reservoir. So, to be on the safe side, coring should be carried out far from the main channel. Keeping this in mind, two promising places were selected as possible coring locations, Location 1 and Location 2, as shown in Figure 5.8.

The selection of these two locations is based on the tendency of sediment to always deposit there and to the apparent absence of bar migration.

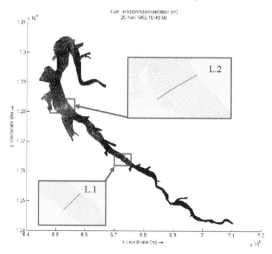

Figure 5.8: Selected coring locations (L1 and L2).

Location 1

Figure 5.9 showed the vertical profiles of deposition at Location 1. The black lines indicate the bed levels along the cross-section at the end of every year: end of 1985 to end of 1992 in Figure 5.9 (a); end of 1992 to end of 2007 in Figure 5.9 (b). In Figure 5.9 (a), the solid lines represent the final bed level of 1985, 1988 and 1991. In Figure 5.9 (b), the solid lines represent the final bed level of 1992, 1995, 1998, 2001, 2004 and 2007. In areas characterized by the absence of bed erosion, the lowest solid line represents the first year, whereas the top line represents the final year of the computation.

At this location, deposition always occurs at the right side of the reservoir; from 0 m to 250 m. Erosion occurs due to channel shift in the middle of the reservoir. The last 200 m at left side of the reservoir, from 1500 m to 1700 m, are again characterized by deposition only.

The dominant deposited sediment in 1986 (dry year) is sand. Sand content is higher in the period 1985-1992 (Figure 5.9 (a)) than in the following 15 years (Figure 5.9 (b)). The general trend in the years 1989, 1990, 1991 and 1992 is deposition of coarser sediment in the deepest area (main channel).

Deposition and stratification are visible at the sides of the reservoir. These areas become dry at the end of the dry season, are always characterized by deposition and therefore promising coring locations.

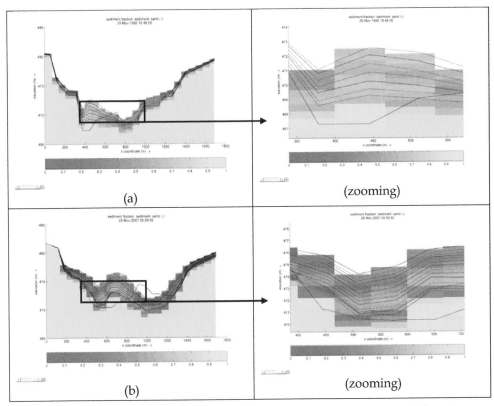

Figure 5.9: Vertical profiles of bed composition at Location 1 seen from downstream. Left: entire cross-section. Right: zoomed areas. (a) period 1985-1992, (b) period 1992-2007.

Location 2

Figure 5.10 showed the vertical profile of deposition at Location 2. Black lines represent the bed levels at the end of every year. The left side of the section, for approximately 3 km, has only been subject to deposition. In this area, the reservoir is relatively wide. Most of sediment deposited in this section is silt with only a minor percentage of sand, as shown in Figure 5.10. Bed level changes in 1985-1992 were shown in Figure 5.10 (a). Figure 5.10 (b) showed the bed level changes in 1993-2007.

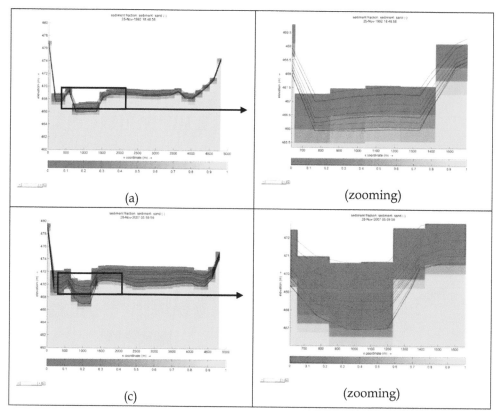

Figure 5.10: Vertical profiles of bed composition at Location 2 seen from downstream. Left: entire cross-section. Right: zoomed areas. (a) period 1985-1992, (b) period 1992-2007.

A subsequent field campaign was carried out in the summer 2012 in the two coring locations identified by the model. In location 2, about 25 km upstream the dam body three trenches were excavated and the forth trench was excavated at location 1, 45 km upstream the dam body. The characteristics of the four trenches are summarized in table 5.1.

Table 5.1: Characteristics of trenches inside Roseires Reservoir.

E (m)	N (m)	Depth (m)	Samples interval (m)	Name	Remarks
650539	1280023	2.5	0.50	Trench 1	Left bank
650539	1280023	4.0	0.25	Trench 2	Left bank
653332	1284514	4.0	0.50	Trench 3	Right bank
673117	1286508	2.5	0.50	Trench 4	Right bank

The field measurement proved that layers are mainly distinguishable from the presence of alternations of sand and silt. The sand and silt content for different layers was shown in Figure 5.11. The recognition of wet years and dry years is possible, as demonstrated in the results.

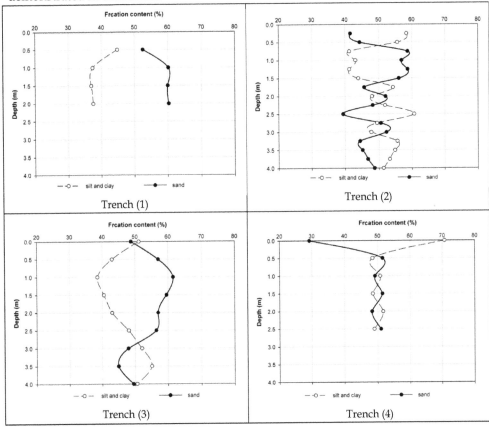

Figure 5.11: Measures sand and silt fraction content in the four trenches.

It was observed that indeed the reservoir, at least in that area, presents soil stratification. The grain size vertical distribution of the soil material, determined at the four trenches, showed that the analysis of D_{90} (sand content) allowed recognizing clear layering (Figure 5.12). The deposition of a low water season and high water season is relatively recognized. This might be due to coarse sand used and the under layer thickness selected in the model.

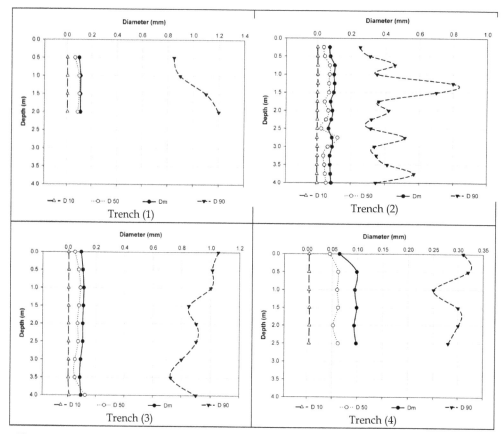

Figure 5.12: Stratification in the four trenches.

Comparison between sand content measured from different layers along the four trenches and the model results are presented in Figure 5.13. The Sand content in the model appeared to be less than the measured and this may be due to the model big grid size and the sediment boundary condition.

In conclusion based on a subsequent field campaign carried out in the summer 2012, it is observed the presence of clear layers at the four selected trenches, but layers are mainly distinguishable from the presence of coarse sand (wet years) rather than from alternations of sand and silt.

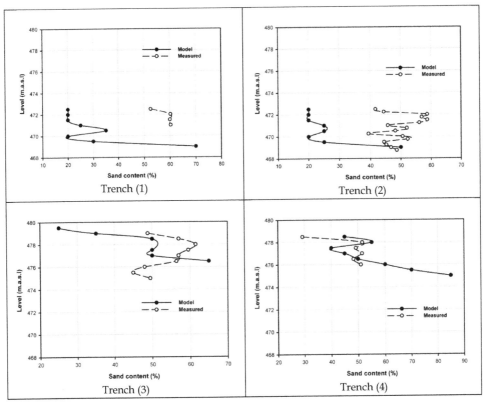

Figure 5.13: Comparison between sand content in model and measurement.

5.3   SEDIMENT FINGERPRINTING

5.3.1   Samples collection

Samples were collected from two zones in the Blue Nile River Basin; zone one is the upper basin where erosion takes place (source) and zone two is Roseires Reservoir where a significant amount of the sediment coming from the upper basin is deposited (sink). In the upper basin (zone one), 28 samples were collected from single locations; with spatial distribution covering almost all eroded area in the whole upper basin including the eroded soil and rocks. In the lower part of the basin, 4 locations inside the flood plain of Roseires Reservoir were selected. Each location was selected based on the outputs of simulation the sedimentation processes inside the reservoir using Delft 3D model allowing to select locations where the reach shows no erosion processes.

In each of the 4 locations trenches were dig up to depth from 2.5 m to 4 m and samples were collected every 50 cm, while in one trench the samples were collected every 25 cm for a total of 35 samples (Table 5.1).

The location of collected sediment samples are shown in Figure 5.14. Photos showing samples collection from Jemma basin eroded area, sample collected for land slide in the way to Didessa and two trenches inside Roseires Reservoir were illustrated in Figure 5.15 for collected samples.

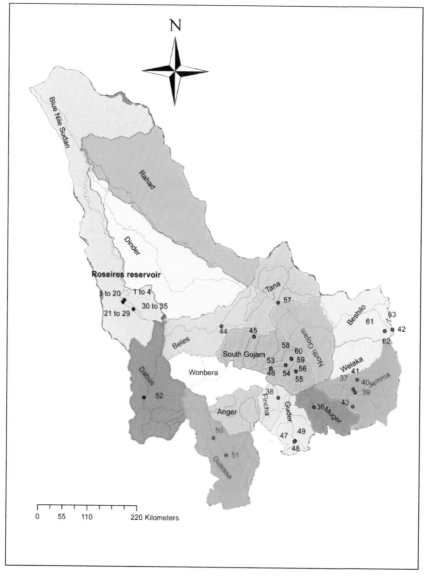

Figure 5.14: The location of collected soil samples, the number as described in detail in table 5.2 and table 5.3.

Figure 5.15: Captured photos showing samples collection from Jemma basin (upper left), Didessa Basin (upper right), trench 3 inside Roseires Reservoir (lower left) and trench 4 inside Roseires Reservoir (lower right).

5.3.2   X-ray Diffraction (XRP)

The samples were dried in an oven at 100 Celsius degree for 24 hours, followed by manual crushing of soil and rock. The crushed samples were further sieved to recover the smaller than 100 μ meter fraction prior to be analyzed by XRD. XRD patterns were recorded in a Bragg-Brentano geometry using a Bruker D5005 diffractometer equipped with Huber incident-beam monochromator and Braun PSD detector. Data collection was carried out at room temperature using monochromatic Cu radiation (Kα1 λ = 0.154056 nm) in the 2θ region between 10° and 90°, step size 0.038 degrees 2θ.

All samples were measured under identical conditions. The samples, of about 20 milligrams each, were deposited on a Si (510) wafer and were rotated during measurement. Data evaluation was done with the Bruker program EVA.   The measured XRD patterns in the 2θ range 10-60 degrees. The outputs are shown in 2θ/count graphs showing different series in different colours; an example is depicted in Figure 5.16. The colour bars give the peak positions and intensities of the identified phases, such as found using the ICDD pdf4 database. All patterns are background-subtracted, meaning the contribution of air scatter and possible fluorescence radiation is subtracted. Possibly present amorphous parts in the samples are not shown.

5.3.3   Cluster analysis

Cluster analysis was applied to compare the six minerals content resulting from XRD analysis for the 63 samples to define "similarity" and a method to quantify how "similar" two data points are. A linking method between clusters and a distance measure are required to be specified on which similarity is based. For the present analyses, linkage was taken as the average of the distances in mineral contents between all pairs in the two groups (known as average linkage) and distance was taken to be Euclidean distance after measurements in each fraction were standardized to have common variation (de Meijer et al., 2001). The average clustering was found to performs better than simple linkage and complete linkage (Cortés et al., 2007). Cluster analysis can be readily carried out in most of the standard statistical packages. The analyses for this study was carried out using the MINITAB software package (Minitab, Inc., 1998) and were based on the square roots of the observations to induce better symmetry and avoid any undue influence of a few large values (de Meijer et al., 2001; Minitab, 1998).

5.3.4   Mineral content results

Example of the XRD pattern of the random powder analysis is shown in Figure 5.16.

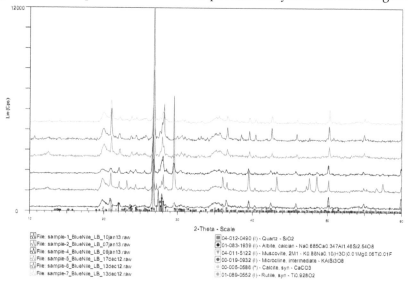

Figure 5.16: X-ray diffraction results-Roseires Reservoir left bank (samples 1-7).

XRD results indicate that the sediment samples from the four trenches inside Roseires Reservoir and eroded soils and rocks samples from the upstream catchment consist of Quartz [SiO_2], Albite [$Na_{0.685}Ca_{0.347}Al_{1.46}Si_{2.54}O_8$], Muscovite 2M1,[$K_{0.86}Na_{0.10}(H_3O)_{0.01}Mg_{0.06}Ti_{0.01}Fe_{0.07}Al_{2.88}Si_{3.02}O_{10}(OH)_2$], Calcite [$CaCO_3$], Rutile [$TiO2$] and Microcline [$KAlSi_3O_8$].
Sample 49 from the upstream catchment consists also Augite [$Ca(Mg,Fe,Al)(Si,Al)_2O_6$]. The weight percentage of mineral content for the upstream samples and the Roseires Reservoir samples are shown in Table 5.2 and Table 5.3 respectively.

Table 5.2: Minerals percentage for the samples collected from four trenches inside Roseires Reservoir

No	Location	Depth*(cm)	Quartz	Albite	Muscovite	Microcline	Calcite	Rutile
1	L.B (trench 1)	50	36	25	32	0	5	3
2	L.B (trench 1)	100	17	14	30	0	39	0
4	L.B (trench 1)	250	37	17	22	5	16	0
5	L.B (trench 2)	25	39	32	30	0	0	0
6	L.B (trench 2)	50	45	18	31	5	0	0
7	L.B (trench 2)	75	35	23	38	4	0	0
9	L.B (trench 2)	125	28	28	38	4	3	0
10	L.B (trench 2)	150	39	10	35	15	0	0
11	L.B (trench 2)	175	42	14	39	2	2	0
13	L.B (trench 2)	225	36	21	34	6	3	0
14	L.B (trench 2)	250	33	22	36	5	3	0
15	L.B (trench 2)	275	13	9	12	0	67	0
16	L.B (trench 2)	300	34	19	39	3	4	0
17	L.B (trench 2)	325	35	25	33	0	7	0
18	L.B (trench 2)	350	43	27	26	0	4	0
19	L.B (trench 2)	375	33	28	38	0	2	0
20	L.B (trench 2)	400	36	31	26	5	2	0
21	R.B (trench 3)	0	34	16	49	0	0	0
22	R.B (trench 3)	50	20	30	28	3	19	0
23	R.B (trench 3)	100	24	24	31	3	22	0
24	R.B (trench 3)	150	21	24	28	4	23	0
25	R.B (trench 3)	200	15	12	28	2	44	0
26	R.B (trench 3)	250	28	11	44	0	17	0
27	R.B (trench 3)	300	27	25	47	0	0	0
28	R.B (trench 3)	350	23	23	54	0	0	0
29	R.B (trench 3)	400	23	24	49	4	0	0
30	R.B (trench 4)	0	38	26	36	0	0	0
31	R.B (trench 4)	50	33	32	34	1	0	0
32	R.B (trench 4)	100	27	34	34	6	0	0
33	R.B (trench 4)	150	20	24	29	3	24	0
34	R.B (trench 4)	200	26	37	28	0	9	0
35	R.B (trench 4)	250	24	25	35	0	17	0

* Depth is measured from the ground surface

L.B: Left bank of Blue Nile River

R.B: Right bank of Blue Nile River

Table 5.3: minerals percentage for the samples collected from the upper basin

Sample	Location	Basin	Quartz	Albite	Muscovite	Microcline	Calcite	Rutile
36	Muger	Muger	9	0	89	0	0	2
37	Jemma R (L.B)	Jemma	2	0	0	0	98	0
38	Fincha R	Fincha	7	93	0	0	0	0
39	Jemma R (R.B)	Jemma	33	27	39	0	0	0
40	Welaka	Welaka	54	0	46	0	0	0
41	Welaka R	Welaka	11	89	0	0	0	0
42	Wallo-Dessie	Jemma	20	17	63	0	0	0
43	Debre Birhan	S. Gojam	43	0	57	0	0	0
44	Beles	Beles	100	0	0	0	0	0
45	Bahridar	N. Gojam	49	5	17	6	23	0
46	Birr	Birr	40	0	54	6	0	0
47	Guder basin	Guder	42	0	0	0	0	0
48	Guder R	Guder	35	0	65	0	0	0
49	Guder Rock	Guder	0	100	0	0	0	0
50	Didessa R	Didessa	23	38	31	8	1	0
51	Didessa R	Didessa	47	40	0	11	2	0
52	Dabus R	Dabus	17	0	83	0	0	0
53	Jedeb Rock	S. Gojam	2	0	98	0	0	0
54	Jedeb soil	S. Gojam	41	0	51	0	7	0
55	Chamoga Rock	S. Gojam	51	0	49	0	0	0
56	Chamoga soil	S. Gojam	51	0	49	0	0	0
57	Abay	Tana	52	0	48	0	0	0
58	Tamamai Guly	S. Gojam	64	0	36	0	0	0
59	Tamamai Rock	S. Gojam	68	0	32	0	0	0
60	Tamamai Guly	S. Gojam	45	0	55	0	0	0
61	South Wello	Jemma	16	9	55	20	0	0
62	South Wello	Jemma	24	9	61	7	0	0
63	South Wello	Jemma	31	0	69	0	0	0

The general characteristics of the resulted mineral are discussed below with the sink and source areas.

The relative chemical stability or susceptibility to chemical decomposition of common minerals is generally known. The Goldich dissolution series is a way of predicting the relative stability or weathering rate of various minerals on the Earth's surface(Goldich, 1938). S. S. Goldich came up with the series in 1938 after studying soil profiles. He found that minerals that form at higher temperatures and pressures are less stable on the surface than minerals that form at lower temperatures and pressures. His diagram shows that the order of weathering series agrees well with the least bond strength in each of the minerals involved. That seems reasonable, because weathering, or the destruction of minerals, requires breaking of bonds in those minerals. He concluded that igneous silicate minerals weather in an order much like that of Bowens Reactions Series, with mafic silicates the most susceptible to weathering and muscovite and quartz the least susceptible (Goldich, 1938). Quartz is the most stable phase at ambient P/T and amorphous silica is the least stable phase, which means quartz has the lowest solubility, however experimental studies showed the solubility of quartz in high temperature and pressure conditions is much higher(Anderson and Burnham, 1965; Fournier and Potter Ii, 1982; Rimstidt, 1997; Wang et al., 2011).

The formation of secondary minerals in soils generally results from the recombination and addition of ions and molecules from the soil solution to the solid phase. For example, Sodium feldspar (albite NaAlSi3O8) and Potassium feldspar (microcline KAlSi3O8) reacts with hydrogen iron and water to form kaolinite (Hefferan and O'Brien, 2010).

$$2NaAlSi_3O_8 + 9H_2O + 2H^+{}_{(aq)} \rightarrow 2Na^+{}_{(aq)} + H_4SiO_4{}_{(aq)} + Al_2Si_2O_5(OH)_4 \qquad\qquad 5.5$$

Albite (Na feldsp)   kaolinite

In addition to that reaction, feldspars are the most abundant mineral group in Earth's crust, ion exchange processes are an important part of chemical decomposition.

$$NaAlSi_3O_8 + H^+{}_{(aq)} \rightarrow HAlSi_3O_8 + Na^+{}_{(aq)} \qquad\qquad 5.6$$

Calcite is a rock-forming mineral with a chemical formula of $CaCO_3$. It is commonly found throughout the world in sedimentary, metamorphic and igneous rocks. In regions of limited rainfall where evapo-transpiration is greater than infiltration rate, $CaCO_3$ is likely to accumulate. The chemical reaction responsible for this process is:

$$Ca^{+2} + H_2O + CO_2 \rightleftharpoons CaCO_3 + 2H^+ \qquad\qquad 5.7$$

5.3.5   Cluster analysis results

All samples from Roseires Reservoir and upper catchments were used in the analyses to determine the overall patterns of similarity between samples from different parts of the catchment. The dendrogram shown in Figure 5.17 illustrates the combined five mineral fractions Quartz, Albite, Muscovite, Microcline and Calcite. However the Rulite mineral was omitted from the analysis due to its appearance in one sample only. The dendrogram visualizes the order in which samples, or groups of samples, combine to form clusters with similar mineral content and the similarity levels at which the combinations of minerals occur. Similarity is a measure of distance between clusters relative to the largest distance between any two individual samples. One hundred percent similarity means the clusters were zero distance apart in their sample measurements.

We observed seven major groupings from A to G based on the similarity occurred between samples; group A contains all samples with similarity of 85% or more. Whereas, group B contains samples that have similarity of 80% with group A. Furthermore, group C represents all samples that have similarity of 75% with subgroup A and B.

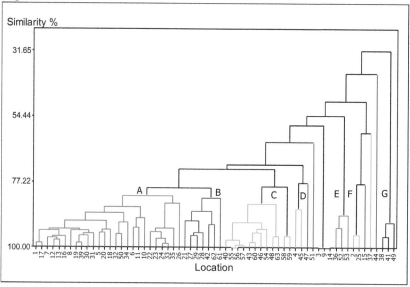

Figure 5.17: Dendrogram for the mineral content with Average Linkage and Euclidean Distance

Based on the dendrogram of cluster analysis for the 63 samples in Figure 5.17, the following interpretation may be suggested:

✓ All samples in group A are similar with a similarity of 85% or more, 24 samples inside Roseires Reservoir are similar to sample 39 from Jemma River and sample 50 from Didessa River. A similar situation occurs in group B since four samples inside Roseires Reservoir are similar to sample 42 from eroded soil near Wallo-Dessie in Jemma basin and sample 61 and sample 62 from eroded soil from Eyasu Ager in South Wello-Gully in Welaka basin. The two subgroups A and B are still similar at more than 80% similarity.

✓ All samples in group C are similar with a similarity of 77% or more, the samples in this group are from the upper basin including sample 40 (Welaka), 43, 54, 55, 56, 58, 59 and 60 (South Gojam), 57 (Tana), 46 (Birr), 48 (Guder) and 63 (Jemma). The similarity between subgroup C and subgroup A and B is less than 75%.

✓ Group D consists of sample 4 from Roseires Reservoir, which similar to sample 45 (North Gojam) and samples 47 (Guder) from the upper basin with similarity of 65% with group (A+B+C).

✓ Group E consists of sample 36, 52 and 53 from Muger, Dabus and South Gojam sub basins respectively has similarity of 55% with group (A+B+C+D).

✓ Group F consists of sample 2, 25, 16 from Roseires Reservoir which similar to sample 37 from Jemma Basin in the upper basin and has similarity of 50% with group (A+B+C+D+E).

Based on the cluster analysis, we suggest that most of the sediment deposited inside Roseires Reservoir is originated by erosion from Jemma, Didessa and South Gojam sub basins in the upper Blue Nile basin. The basin is characterized by a large variability in topography, natural vegetation cover, rainfall, temperature, soils, and lithology that explain a large part of the observed variability in sediment yield. Topography is often indicated as an important factor controlling erosion rates (Tamene et al., 2006). Climatic factors such as rainfall depths and intensities and air temperature are regularly cited as important factors controlling riverine sediment fluxes (Nyssen et al., 2008). Vegetation cover is reported to have significant importance on sediment fluxes. Although vegetation density is controlled by anthropogenic activity, it may be assumed that broad regional patterns in vegetation cover are strongly naturally controlled and that dense vegetation can act as a biophysical barrier to soil erosion (de Vente et al., 2011; Kettner et al., 2010). Based on sediment yield data for 11 catchments in northern Ethiopia, Tamene (2006), concluded that no significant relation was found between sediment yield and the proportion of dense vegetation, cultivated land, bush/shrub cover or bare land.

It is reported in Abay master plan that high soil erosion rates are apparent from the most important agricultural lands, especially those of Gojam. Developing agriculture in Didessa, Jemma and Illubabor may be expected to promote similar loss rates (BCEOM, 1999).

The Leptosols mostly occurring in Jemma sub basin are shallow soils with limited profile development and are prone to drought. They occur on steep slopes, they are exposed to a high degree of erosion. The Leptosols in Jemma sub basin are lying above deep Cambisols and cultivated in altitude between 2300 and 3500 m for about 130-230 days/year, divided between two rainy seasons. In upper Didessa, the Alisols are deeply leached soils with only moderate to low fertility. Fertility tends to be highest in the topsoils, meaning that erosion has deep and permanent impact on these soils. Nitisols and Acrisols are similar in general respects to Alisols. However, Nitisols are in principle better soils, while Acrisols are the most leached with the lowest pH. More detailed survey work would assist in separating such soils, and directing development away from the Acrisols.

Cluster analysis was performed for different layers of deposited sediment inside Roseires Reservoir. Two samples were collected from the reservoir at depth of 4 m below the surface including trench 2 and trench 3 and other two samples were collected at a depth of 3 m from the same trenches. Delft3D model showed that the sediment layers at 4 m depth and 3 m depth were deposited approximately in the period 1986 to 1991 and 1993 to 1997 respectively. Table 5.4 provides minerals content for the samples collected from 4 m (no 1 and 2 up) and 3 m (no 1 and 2 down) depth inside the reservoir and samples collected from the upper basin (no 3 to 30).

Table 5.4: minerals percentage for the samples collected from 4m and 3 m depth inside Roseires Reservoir and the upper basin

No	Location	Basin	Quartz	Albite	Muscovite	Microcline	Calcite	Rutile
1	Roseires Dam	Trench 2(4 m)	36	31	26	5	2	0
2	Roseires Dam	Trench 3(4 m)	23	24	49	4	0	0
1	Roseires Dam	Trench 2(3 m)	34	19	39	3	4	0
2	Roseires Dam	Trench 3(3 m)	27	25	47	0	0	0
3	Muger	Muger	9	0	89	0	0	2
4	Jemma R (L.B)	Jemma	2	0	0	0	98	0
5	Fincha R	Fincha	7	93	0	0	0	0
6	Jemma R (R.B)	Jemma	33	27	39	0	0	0
7	Welaka	Welaka	54	0	46	0	0	0
8	Welaka R	Welaka	11	89	0	0	0	0
9	Wallo-Dessie	Jemma	20	17	63	0	0	0
10	Debre Birhan	S. Gojam	43	0	57	0	0	0
11	Beles	Beles	100	0	0	0	0	0
12	Bahridar	N. Gojam	49	5	17	6	23	0
13	Birr	Birr	40	0	54	6	0	0
14	Guder basin	Guder	42	0	0	0	0	0
15	Guder R	Guder	35	0	65	0	0	0
16	Guder Rock	Guder	0	100	0	0	0	0
17	Didessa R	Didessa	23	38	31	8	1	0
18	Didessa R	Didessa	47	40	0	11	2	0
19	Dabus R	Dabus	17	0	83	0	0	0
20	Jedeb Rock	S. Gojam	2	0	98	0	0	0
21	Jedeb soil	S. Gojam	41	0	51	0	7	0
22	Chamoga Rock	S. Gojam	51	0	49	0	0	0
23	Chamoga soil	S. Gojam	51	0	49	0	0	0
24	Abay	Tana	52	0	48	0	0	0
25	Tamamai Guly	S. Gojam	64	0	36	0	0	0
26	Tamamai Rock	S. Gojam	68	0	32	0	0	0
27	Tamamai Guly	S. Gojam	45	0	55	0	0	0
28	South Wello	Jemma	16	9	55	20	0	0
29	South Wello	Jemma	24	9	61	7	0	0
30	South Wello	Jemma	31	0	69	0	0	0

The cluster analysis showed that sample 1 from Roseires Reservoir at depth of 4 m is similar to sample 6 from Jemma River and sample 17 Didessa River and having similarity of more than 90%. Sample 2 from the reservoir at depth of 4 m is similar to Sample 9, 28 and 29 from South Wello in Jemma Basin with similarity of about 90%. The results are shown is Figure 5.18 (left)

The cluster analysis showed that sample 1 and sample 2 from Roseires Reservoir at depth of 3 m (Figure 5.18 right) are similar to sample 6 from Jemma River and sample 17 Didessa River and having similarity of more than 85%

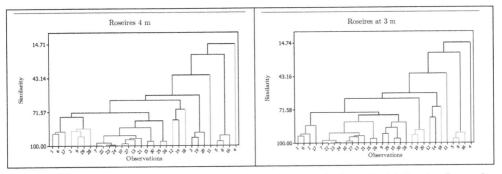

Figure 5.18: Cluster analysis for samples collected from the upper basin and inside Roseires Reservoir at depth of 4 m (left) and 3 m (right).

Four samples were collected from trench 1, trench 2, trench 3 and trench 4 at depth of 2 m below the surface as well as other four samples at a depth of 1 m from the same trenches. Delft3D model showed that the sediment layers at 2 m depth and 1 m depth were deposited approximately in the period 1998 to 2002 and 2003 to 2007 respectively. Table 5.5 provides minerals content for the samples collected from 2 m (no 1 and 4 up) and 1 m (no 1 and 4 down) depth inside the reservoir and samples collected from the upper basin (no 5 to 32).

Table 5.5: minerals percentage for the samples collected from 2 m and 1 m depth inside Roseires Reservoir and the upper basin.

No	Location	Basin	Quartz	Albite	Muscovite	Microcline	Calcite	Rutile
1	Roseires Dam	Trench 1(2 m)	37	17	22	5	16	0
2	Roseires Dam	Trench 2(2 m)	36	21	34	6	3	0
3	Roseires Dam	Trench 3(2 m)	15	12	28	2	44	0
4	Roseires Dam	Trench 4(2 m)	26	37	28	0	9	0
1	Roseires Dam	Trench 1(1 m)	17	14	30	0	39	0
2	Roseires Dam	Trench 2(1 m)	28	28	38	4	3	0
3	Roseires Dam	Trench 3(1 m)	24	24	31	3	22	0
4	Roseires Dam	Trench 4(1 m)	27	34	34	6	0	0
5	Muger	Muger	9	0	89	0	0	2
6	Jemma R (L.B)	Jemma	2	0	0	0	98	0
7	Fincha R	Fincha	7	93	0	0	0	0
8	Jemma R (R.B)	Jemma	33	27	39	0	0	0
9	Welaka	Welaka	54	0	46	0	0	0
10	Welaka R	Welaka	11	89	0	0	0	0
11	Wallo-Dessie	Jemma	20	17	63	0	0	0
12	Debre Birhan	S. Gojam	43	0	57	0	0	0
13	Beles	Beles	100	0	0	0	0	0
14	Bahridar	N. Gojam	49	5	17	6	23	0
15	Birr	Birr	40	0	54	6	0	0
16	Guder basin	Guder	42	0	0	0	0	0
17	Guder R	Guder	35	0	65	0	0	0
18	Guder Rock	Guder	0	100	0	0	0	0
19	Didessa R	Didessa	23	38	31	8	1	0
20	Didessa R	Didessa	47	40	0	11	2	0
21	Dabus R	Dabus	17	0	83	0	0	0
22	Jedeb Rock	S. Gojam	2	0	98	0	0	0
23	Jedeb soil	S. Gojam	41	0	51	0	7	0
24	Chamoga Rock	S. Gojam	51	0	49	0	0	0
25	Chamoga soil	S. Gojam	51	0	49	0	0	0
26	Abay	Tana	52	0	48	0	0	0
27	Tamamai Guly	S. Gojam	64	0	36	0	0	0
28	Tamamai Rock	S. Gojam	68	0	32	0	0	0
29	Tamamai Guly	S. Gojam	45	0	55	0	0	0
30	South Wello	Jemma	16	9	55	20	0	0
31	South Wello	Jemma	24	9	61	7	0	0
32	South Wello	Jemma	31	0	69	0	0	0

The results of cluster analysis for 4 samples from Roseires Reservoir at depth of 2 m and 28 samples from upper basin (Figure 5.19 left) are explained below:

A. Sample 1 is similar to sample 14 from Bahridar at North Gojam sub basin and having similarity of more than 90%, these two samples are similar to sample 16 from Guder sub basin with similarity of about 80%.
B. Sample 2 is similar to sample 8 from Jemma River with similarity of about 95%.
C. Sample 4 is similar to sample 19 from Didessa River with similarity of about 95%.
D. Group B is similar Group C with more that 85% similarity
E. Group A is similar Group D with more that 70% similarity
F. Sample 3 showed similarity of about 55% to most of samples from the upper basin

The results of cluster analysis for 4 samples from Roseires Reservoir at depth of 1 m and 28 samples from upper basin (Figure 5.19 right) are explained below:
A. Sample 1 is similar to sample 3 from Roseires Reservoir and having similarity of more than 80%
B. Sample 2 is similar to sample 8 from Jemma River and having similarity of more than 95%
C. sample 4 and sample 19 Didessa River and having similarity of more than 95%
D. Group B and Group C have a similarity of more that 90%.
E. Group D and Group A have a similarity of more that 75%.

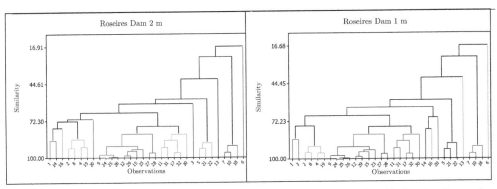

Figure 5.19: Cluster analysis for samples collected from the upper basin and inside Roseires Reservoir at depth of 2 m (left) and 1 m (right).

5.4 CONCLUDING REMARKS

The intervals between the measured cross-sections in bathymetric surveys of Roseires Reservoir of 2009 are 2 to 5 km from each-other and more than 5 km in the bathymetric surveys of 1985, 1992 and 2007. This may create inaccuracy to prepare the reservoir bed topography in Delft3D software considering the length of Roseires

Reservoir (80 km) and its meandering shape. Moreover, the 1985 survey misses seven sections. This, together with the unavailability of data regarding the river bed topography upstream of Famaka in 1985, 1992 and 2007, creates not only model inaccuracy, but also inaccuracy in the analysis of data, in general.

In the numerical model, the grid-cell size was between 25 to 250 m. Using large grid-cell sizes allowed reduced computational times, but yielded inaccuracies, for instance in the interpolation process to generate bed topographies and in the representation of the natural meandering channel excavated through sediment deposits. Morphodynamic computations were speeded up by using a "morphological factor". However, using a morphological factor of 30.4 to simulate a day hydrodynamically to represent a month morphodynamically creates a water balance problem in the reservoir, since a day is not enough to store the water which is stored in a month and vice-versa in the drawdown period. This problem was reduced by adding and subtracting water to the model as rainfall, or evaporation

Nevertheless, we can use the minerology results for semi quantitative distribution of sediment provenance. The outputs of this study highlight the most critical sub basins for sediment input into the Blue Nile system.

Cluster analysis proved to be a useful semi-quantitative technique for analyzing the data and determining linkages between sediment deposited inside Roseires Reservoir and its source in the Upper Blue Nile Basin.

Spatial sampling resolution was one of the main limitations, since it is impossible to sample the whole basin due to time, logistics and accessibility constraints (mountain topography and no road access to most areas).

The XRD method like all analytical techniques has some shortcomings. The diffraction pattern is unique, in practice there are sufficient similarities between patterns as to cause confusion. There are probably in excess of two million possible unique "phases", of which only around 120 000 on the reference database in the International Centre for Diffraction Data (ICDD file) as single-phase patterns exist. Moreover, the diffraction method is not comparable in sensitivity to the other X-ray-based techniques. Whereas in X-ray spectrometry one can obtain detection limits in the low parts per million regions, the powder method has difficulty in identifying several tenths of one percent. To this extent it is less sensitive than the fluorescence method by about three orders of magnitude (Ascaso and Blasco, 2012; Jenkins, 2000).

Extensive searching has not revealed weathering studies specific to the Blue Nile Basin.

Chapter 6
MORPHODYNAMICS IMPACTS OF GRAND ETHIOPIAN RENAISSANCE DAM AND ROSEIRES HEIGHTENING ALONG BLUE NILE RIVER

Summary
In this chapter, the sediment transport along the entire Blue Nile River was studied using the modelling system SOBEK-River/Morphology to perform morphological simulations for the Blue Nile River network system. The model was calibrated and validated and further used to assess the morphological impacts of the planned developments, including Roseires Dam heightening and the construction of the Grand Ethiopia Renaissance Dam (GERD).

6.1   BACKGROUND

The construction of dams and reservoirs represents a great achievement for the management of the water resource, but at the same time it creates a relevant disturbance to the river ecology and morphology. Nevertheless, reservoirs are built for a number of reasons, , such as hydropower, irrigation, water supply, flood mitigation (Rãdoane and Rãdoane, 2005). Unfortunately, reservoirs have a limited life, mainly due to sedimentation. Sedimentation not only decreases the life span of the reservoir, but it also changes the river morphology in the downstream and upstream parts of the river.

The world's reservoirs are presently being filled up with sediments at a rate of approximately 1% per year, though some countries have a higher percentage of reservoir sedimentation, such as China (about 2.3%)(Hu et al., 2010; White, 2001). This implies that within 50 years the water storage in reservoirs will be half of the current storage. This will have huge economic and environmental consequences, especially for semi-arid environments where a lot of reservoirs have been constructed. In addition, sediment deposition can have huge implications on the ecosystem and the coastal development downstream of the dam (de Vente et al., 2005). Therefore, it is vital to study and predict the sediment yield at the basin scale and realize which factors determine the sedimentation rates of reservoirs. This knowledge allows us to take effective measures against reservoir sedimentation, water shortage and erosion of coast and river banks.

Blue Nile River is an alluvial river with rich silt (Gismalla, 2009). Alluvial river channels are highly dynamic in nature as they pass through alluvium, which are loose sedimentary materials and are formed through entrainment, transportation and deposition throughout the channel (Church and Rood, 1983) as reported in (Rosgen, 1994). They involve complex phenomena like turbulent flow, secondary flow, sediment transport, bank erosion process and so on.

The human-induced change on river systems by building dams has major effects both upstream and downstream the dam (Brandt, 2000). Apart from the interception of sediments by the reservoir, the other major cause of morphological changes below dams is the substantial reduction of water flow transport capacity from the regulation of the natural runoff with a virtual disappearance of large and, especially, medium floods. The impact of dams on river hydrology include changes in the timing, magnitude and frequency of low and high flows (Magilligan and Nislow, 2005; Yuan et al., 2012), the trapping of sediments by the reservoirs may give course to a number of negative consequences: besides the reduction of the storage capacity, reservoir sedimentation produces important modifications of river morphology upstream and downstream of the dams (Dai and Liu, 2013; Gupta et al., 2012; Li et al., 2011).

The main objective of this chapter is to study the sediment transport along the entire Blue Nile River and secondly, to assess the morphological impacts of the planned developments, including Roseires Dam heightening and the construction of the Grand Ethiopia Renaissance Dam (GERD) using Sobek River/Morphology model.

6.2   NEW DEVELOPMENT

The Blue Nile River shown Figure 2.2 b, is experiencing intensive new development of dam construction across the main river, both in Ethiopia and in Sudan. The Grand Ethiopia Renaissance Dam (GERD) is currently under construction and the Roseires Dam has been heightened by 10 m. The Roseires Reservoir commenced operation in January 2013. The study area consists of the Blue Nile River reach upstream of Roseires dam.

6.2.1   Roseires Dam heightening

The operation of Roseires Dam after heightening was studied by SMEC (2012), two models approaches were performed to study the reservoir filling after the heightening. Spreadsheet model was developed to model the second filling of Roseires Reservoir at 10 days time step. The model indicated that staring the second filling on 11 September each year and filling at uniform rate over 40 days, is likely to be a good compromise between maximizing reliability of supply for irrigation while minimizing sediment trapping by the reservoir. The results of the filling on 11 September days were confirmed using HEC Ressim model. The result of the compromise operation is shown in Figure 6.1.

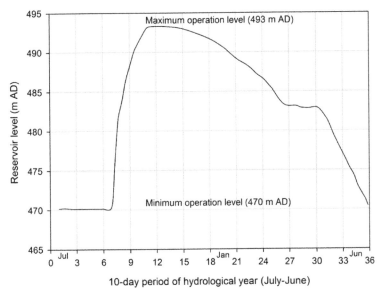

Figure 6.1: Roseires Reservoir operation after heightening(Data obtained from SMEC 2010).

6.2.2   Grand Ethiopian Renaissance Dam (GERD)

Grand Ethiopian Renaissance Dam (under construction) is located approximately 40 km downstream of the confluence with the River Beles at a narrow point about 30 km upstream of the Ethiopian-Sudanese border with a catchment area of some 172,250 km². At this point the river bed is at elevation 500 m (AD), i.e., about 10 m above what the elevation of the Roseires dam once after heightened (Salini and SP, 2010).

Five hydro-power plants were proposed by the Ethiopian from Tana Lake to the Ethiopian-Sudanese border (total gross head = 1290 m) namely GERD, Karadoby, Mendaia, Mabil and Jemma respectively (Salini and SP, 2010).

GERD has been selected as the first priority project because, it is the most economically attractive to produce 15,130 GWh annually and it is the simplest to construct due to the favourable morphology, it also offers the opportunity of opening multiple simultaneous and fully independent construction fronts (Salini and SP, 2010). The general characteristics of GERD are summarised in table 6.1

Table 6.1: Charteristics of Grand Ethiopian Renaissance Dam.

Project section	Parameter	value	unit
Reservoir	Catchment area	172250	Km²
	Full storage level (F.S.L)	640	m.a.s.l
	Minimum operation level (M.O.L)	590	m.a.s.l
	Total storage volume	63350	Mm³
	Live storage volume	51600	Mm³
	Dead storage volume	11750	Mm³
	Annual average runoff	1547	m³/s
Main dam	Type	RCC	-
	Height	145	m
	Crest length	1780	m
	Crest elevation	644	m.a.s.l
Saddle dam	Type	CCRD	-
	Height	45	m
	Crest length	4800	m
	Crest elevation	644	m.a.s.l
Left powerhouse	Type	outdoor	-
	units	5/Francis	No/type
	Installed power	5x350	MW
Right powerhouse	Type	outdoor	-
	units	10/Francis	No/type
	Installed power	10x350	MW
Energy	Max net head	133	M
	Total installed power	5250	MW
	Average power production	15128	GWh/y

6.3 MODEL DESCRIPTION

The modelling system SOBEK-River/Morphology (Hauschild and Sloff, 2009) was used to perform morphological simulations for the Blue Nile River network system. The numerical model is the coupled system of Sobek-River; module of Sobek-Rural code that allows hydrodynamic computations on the large scale, with the DelWAQ code which allows for simulating the morpholological processes including the fine sediment. Both programs run together and exchange data via DelftIO routines concerning updating of cross-sections and of the hydrodynamic parameters such as velocity and water level distribution (Figure 6.2).

The model simulates uniform and graded sediment. For the uniform sediment approach, the time-dependent development of the bed level can be simulated, while in case of the graded sediment approach the development of the bed composition will be accounted for as well (Hauschild and Sloff, 2009).

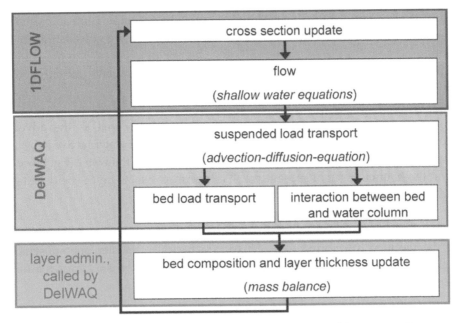

Figure 6.2: SOBEK-River/Morphology computation process (Source; model user manual).

6.3.1   Sobek-River model

The first component of the coupled Sobek -River/Morphology model is Sobek-River (module of Sobek-Rural) that used to solve the hydrodynamics equations and further as input for DelWAQ model. The flow in one dimension is described by the continuity equation and the momentum equation (chapter 4).

In Sobek, the simulation of hydrodynamic module should be successful before proceeding to the sediment transport module. Sobek simulate the hydrodynamics with implicit models. The Courant condition for stability does not have to be fulfilled in order to assure the stability of the calculations in implicit models, but it will influence the accuracy of the solution.

6.3.2   DelWAQ model

DelWAQ is water quality model framework. It solves the advection diffusion-reaction equation on a predefined computational grid and for a wide range of model substances. The model allows great flexibility in the substances to be modeled, as well as in the processes to be considered. DelWAQ is not a hydrodynamic model, so information on flow fields will be derived from the Sobek River model.

DelWAQ process library

Water quality model consists of many components such as substances, processes, items and fluxes. The substances are all constituents for which a concentration or density can be computed by the water quality model e.g.; temperature and sediment. Essentially the substances and processes for morphology are considered for two phases:

1. Sediment transported in the water column (suspended sediment concentration).
2. Sediment available in the bed (bed-load and bed layers).

The concentration/density is influenced by processes as well as the advection diffusion equation (ADE) if the substance is carried with the water (active substance) or without the ADE for substances that are not carried with the water (e.g. those that lay on the bottom, inactive substances). Each process has its input coefficients and produces one or more fluxes.

Items are the input variables for the processes, but also the output variables (that may be input to other processes). Fluxes are basically transfer one substance to another. In terms of morphology, e.g. the process of resuspension creates a flux of sediment from the river bed into the water column.

The DelWAQ grid

The water quality module uses a separate grid called a segment that represents a small piece of the river containing a volume of water and sediment. The segments are connected with each other by contact areas called exchanges to allow the exchange of water and sediment via these exchanges. The DelWAQ-segments are defined between two neighboring water level points from the flow grid (Figure 6.3).DelWAQ defines an exchange by the numbers of the two computational cells that share the surface area. This information is written into a so-called pointer-table during pre-processing. The last segment before a model boundary is connected with a virtual segment outside the boundary, which gets a negative segment-ID.

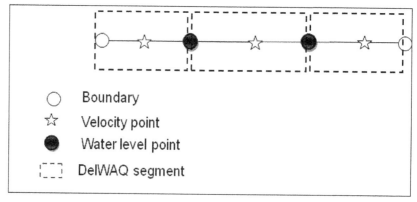

Figure 6.3: *Scheme of the grids of SOBEK-River and DelWAQ, with simple reach, nodes and calculation points (water level points, velocity points and DelWAQ segments).*

bedload transport

The transport of non-cohesive sediment (all types of sand) has been implemented in a slightly different way than most other substances. The module assumes that the transport rate of the non-cohesive sediment is determined by the local instantaneous flow condition. This assumption is commonly assumed for bed load transport. The mass balance of sandy sediments is expressed as the bed level change rate:

$$\frac{\partial z_b}{\partial t} = \frac{1}{1-\theta}\left(\frac{\Delta S_x}{\Delta x} + \frac{\Delta S_y}{\Delta y}\right) \hspace{4cm} 6.1$$

Where: z_b is bed level in m, θ is porosity of the bed and S is sediment transport rate in m²/s.

Suspended-sediment transport

In DelWAQ the advection-diffusion equation is applied to the substances which are suspended in the water-column (not the material that is in the bed layer). In a one dimensional cross- sectional averaged model this equation must be integrated. Three approaches are implemented in Delwaq:

  ✓ For mud/clay fractions the Krone-Partheniades approach is used and currently available in DelWAQ for computing the transport of inorganic matter.

  ✓ For sand-mud fractions an adaptation to the Krone-Partheniades approach is implemented in which the entrainment and deposition terms are modified according to Van Ledden (2003).

  ✓ For (suspended) sand fraction the Galappatti approach is used, in which the entrainment and deposition terms are replaced by Galappatti's source term.

Krone and Parheniades model

The 1D advection-diffusion process of suspended sediment is average mass-balance equation for sediment concentration:

$$\frac{\partial h c_w}{\partial t} + \frac{\partial h u c_w}{\partial t} = \left[w_s c_w + \varepsilon_s \frac{\partial c_w}{\partial z}\right]_{bed} \hspace{3cm} 6.2$$

Where: c_w is the concentration of suspended matter in the water column. The right-hand term in this equation represents the net diffusive and convective flux of sediment at the lower (bed) boundary of the suspension layer. It represents entrainment and deposition, or settling and resuspension fluxes. Furthermore, remaining dispersion terms have been neglected in this equation. Following the equation above the mass balances for particulate (suspended) matter in the water column (cw) can be written following equation.

$$\frac{\partial c_w}{\partial t} = loads + transport - settling + resuspension \hspace{2cm} 6.3$$

The transport, settling and resuspension fluxes can be calculated using the formulations of Krone (1962) and Partheniades (1962). In this concept, the bottom shear stress plays an essential role in defining whether or not sedimentation of suspended particles or erosion of bed material will occur. Sedimentation takes place when the bottom shear drops below a critical value. On the other hand resuspension occurs when the bottom shear exceeds a critical value. Equilibrium between deposition and resuspension if the bed shear stress is in between the critical shear stresses for deposition and resuspension.

A characteristic feature of fine sediments is the ability to form aggregates of flocs that settle to the bottom, depending upon its size, concentration, salinity and turbulence and the chemical conditions of the surrounding water system (Mehta et al., 1989). Turbulence is an important parameter because it affects the flocculation and therefore the settling velocity in two opposing ways.

The rate of downward mass transport (deposition) is equal to the product of the near-bed velocity, the concentration and the probability that a settling particle becomes attached to the bed. Krone (1962) performed flume studies and found that the settling velocity increases with sediment concentration and proposed the following formula:

$$D = w_s . c . \left(1 - \frac{\tau_b}{\tau_d}\right)$$

6.4

Where: D is deposition flux of suspended matter in $g/m^2/d$, w_s is settling velocity of suspended matter in m/d, c is concentration of suspended matter near the bed in g/m^3, τ_b is bottom shear stress (following from a separate process) in Pa and τ_d is critical shear stress for deposition in Pa.

The erosion is directly proportional to the excess of the applied shear stress over the critical erosive bottom shear stress. The formula for erosion of homogeneous beds is based on Partheniades (1962). The erosion flux is limited by the available amount of sediment on the bed. The rate of erosion fluxes of bed material is given by

$$E = P_{m,b} . M . \left(\frac{\tau_b}{\tau_{e,c}} - 1\right)$$

6.5

Where: E is erosion rate in $g/m^2/d$, M is first order erosion rate or erosion parameter (defined by user) in $g/m^2/d$, $P_{m,b}$ is fraction of sediment available on the bed, τ_b is bed shear stress (following from a separate process) in Pa and τ_e is critical shear stress for erosion in Pa.

Sand-mud modelling in Delwaq
In a multi-fraction approach the presence of sand affects the resuspension of clay from the bed layer. The interaction between sand and mud highly depends on clay content. The modelling approach implemented in DelWAQ morphology is that of Van Ledden (2003); which is an extension of the Partheniades-Krone approach for

application to both the sand and mud fractions, accounting for their mutual interactions during erosion. The adaptations involve a revision of the erosion function (above equation). When the mud content $P_{m,b}$ is higher than the critical mud content P_{cr}, the erosion fluxes for both sand (E_s) and mud (E_m) are given by:

$$E_m = P_{m,b}. M. \left(\frac{\tau_b}{\tau_{e,c}} - 1\right) E_m \geq 0 \qquad\qquad 6.6$$

$$E_s = (1 - P_{m,b}). M. \left(\frac{\tau_b}{\tau_{e,c}} - 1\right) E_s \geq 0 \qquad\qquad 6.7$$

Where: E_s, E_m are erosion rate for sand and mud respectively in g/m²/d, M is first order erosion rate or erosion parameter in g/m²/d, $P_{m,\,b}$ is mud content in bed layer, τ_b is bed shear stress (following from a separate process) in Pa and $\tau_{e,\,c}$ is critical shear stress for erosion in cohesive regions in Pa.

When the mud content M_{ob} is lower than the critical mud content P_{cr}, the erosion flux for mud is calculated as follows:

$$E_m = P_{m,b}. M. \left(\frac{\tau_b}{\tau_{e,nc}} - 1\right) E_m \geq 0 \qquad\qquad 6.8$$

Where: $\tau_{e,nc}$ is critical shear stress for erosion in non-cohesive sand-mud mixtures (Pa)

The rate of downward mass transport (deposition) is calculated for mud using the following formula (Krone 1962):

$$D_m = w_m. c_m. \left(1 - \frac{\tau_b}{\tau_d}\right) \qquad\qquad 6.9$$

Where: τ_d is critical shear stress for mud deposition in N/m², c_m is mud concentration and w_m is settling velocity for mud in m/s.

6.3.3   Numerical solution of the advection-diffusion processes

The numerical scheme used in Delwaq is Implicit Upwind scheme with an iterative solver. It starts with a guess for the concentration vector of the previous time-step, and subsequently checks if this guess was correct or not. When the residual is small enough, convergence is assumed and the iterative process stops. Although convergence is guaranteed, still many iterations may be required for convergence.

The mass balance equation used in Delft3D DelWAQ solves advection-diffusion-reaction equation for each computational cell and for each state variable. The simplified representation of the advection-diffusion-reaction equation is given bellow.

$$M_i^{t+\Delta t} = M_i^t + \Delta t \times \left(\frac{\Delta M}{\Delta t}\right)_{Tr} + \Delta t \times \left(\frac{\Delta M}{\Delta t}\right)_P + \Delta t \times \left(\frac{\Delta M}{\Delta t}\right)_S \qquad 6.10$$

Where: M_i^t is the mass at the beginning of a time step, $M_i^{t+\Delta t}$ the mass at the end of a time step, $(\Delta M/\Delta t)_{Tr}$ changes by transport, $(\Delta M/\Delta t)_P$ changes by physical, (bio)chemical or biological processes and $(\Delta M/\Delta t)_S$ changes by sources (e.g. waste loads, river discharges).

Changes by transport include both advective and dispersive transport transport by flowing water and the transport as a result of concentration differences respectively. Changes by processes include physical processes such as settling, (bio)chemical processes such as adsorption and denitrification and biological processes such as primary production and predation on phytoplankton. A special type of processes deals with settling in a 3-dimensional situation, as these processes transport particulate matter from one computational cell to the one below.

Changes by sources include the addition of mass by waste loads and the extraction of mass by intakes. Mass entering over the model boundaries can be considered a source as well. The water flowing into or out of the modelled area over the model boundaries is derived from the hydrodynamic model.

The advection-diffusion equation is solved on the numerical grid of DelWAQ. The advective transport across an exchange can be given as:

$$T_{x_0}^A = v_{x_0} \times A \times C_{x_0} \qquad\qquad 6.11$$

Where: $T_{x_0}^A$ is advective transport at x = x$_0$ in g/s, v_{x_0} is velocity at x = x$_0$ in m/s, A is surface area at x = x$_0$ in m^2 and C_{x_0} is concentration at x = x$_0$ in g/m^3

It is assumed that velocities and concentrations are an average representative value for the whole surface. The dispersive transport across an exchange is assumed to be proportional to the concentration gradient and to the surface area:

$$T_{x_0}^D = -D_{x_0} \times A \times \left.\frac{\partial C}{\partial x}\right|_{x=x_0} \qquad\qquad 6.12$$

Where: T_{xo}^D is dispersive transport at x = x$_0$ in g/s, D_{x_0} is dispersion coefficient at x = x$_0$ in m^2/s, A is surface area at x = x$_0$ (m^2) and $(\Delta C/\Delta x)_{x=xo}$ is concentration gradient at x = x$_0$ in g/m^4.

The minus sign originates from the fact that dispersion causes net transport from higher to lower concentrations, so in the opposite direction of the concentration gradient. The concentration gradient is the difference of concentrations per unit length, over a very small distance across the cross section:

$$\left.\frac{\partial C}{\partial x}\right|_x = lim_{\Delta x \to 0} \frac{C_{x+0.5\Delta x} - C_{x-0.5\Delta x}}{\Delta x} \qquad\qquad 6.13$$

If the advective and dispersive terms are added and the terms at a second surface are included, the one dimensional equation results will be as given below:

$$M_i^{t+\Delta t} = M_i^t + \Delta t \times \left(v_{x_0}C_{x_0} - v_{x_0+\Delta x}C_{x_0+\Delta x} - D_{x_0}\frac{\partial c}{\partial x}\Big|_{x_0} + D_{x_0+\Delta x}\frac{\partial c}{\partial x}\Big|_{x_0+\Delta x}\right) \times A \qquad 6.14$$

Where: M_i^t is mass in volume i at time t in g, Δt is time step in s, $(\Delta C/\Delta x)_{x_0}$ is concentration gradient at $x = x_0$ in g/m^4, A is surface area at $x = x_0$ in m^2, v_{x_0} is velocity at $x = x_0$ in m/s and C_{x_0} is concentration at $x = x_0$ in g/m^3

If the previous equation is divided by the volume V (= $\Delta x\, \Delta y\, \Delta z$), and the time span Δt, then the following equation results in one dimension (Here it is required that all segments have an equal volume).

$$\frac{c_i^{t+\Delta t}-c_i^t}{\Delta t} = \frac{D_{x_0+\Delta x}\frac{\partial c}{\partial x}\big|_{x_0+\Delta x} - D_{x_0}\frac{\partial c}{\partial x}\big|_{x_0}}{\Delta x} + \left(\frac{v_{x_0}C_{x_0}-v_{x_0+\Delta x}C_{x_0+\Delta x}}{\Delta x}\right) \qquad 6.15$$

Taking the asymptotic limit $\Delta t \to 0$ and $\Delta x \to 0$, the advection-diffusion equation for one dimension results:

$$\frac{\partial c}{\partial t} = \frac{\partial}{\partial x}\left(D\frac{\partial c}{\partial x}\right) - \frac{\partial}{\partial x}(vC) \qquad 6.16$$

The transport of bed material i.e. material in the active layer follows from transport formula. The transport rates on "Exchanges" can be considered as sediment fluxes affecting the volume of the considered sediment fraction in the bed layer. The selected numerical scheme for SOBEK should satisfy the requirements for numerical stability, accuracy and robustness.

The scheme has been generalized more by adding upwind components in the discretization of the fluxes. The bed-level changes and changes in bed composition for time step Δt are computed from the mass balance of sediment of bed layer S. The procedure is as follows:

1. Compute erosion/deposition terms from suspended load simulation per fraction
2. Add net bed-load/total-load sediment flux over segment per fraction (multiply the transport capacity with the available volume percentage to get the actual transport rate)
3. Add the lateral sediment per fraction (e.g., dredging). Adding all the fluxes for all fractions, a net change of mass volume of the segment can be computed using the proposed space-discretization. Based on the bed-level change, combined with eventual change in thickness of the mixing layer, the exchange flux between the active layer S and the under layers can be defined.

6.4   DATA ANALYSIS

The flow of the Blue Nile River is characterized by two different periods according to the seasonality of rainfall over the Ethiopian Highlands. The low season extends from

November to June, whereas the flood period takes place between July and October, with maximum flow in August-September. During the wet season, the sediment load is high with a large percentage of fine sediments.

Wash load causes a large amount of fine sediment deposited in the reservoirs. The transport formula cannot calculate wash load, because they compute the sediment transport capacity. Delwaq can simulate both the wash load and the bed load, for this reason the combined model Sobek Rural Delwaq was used to simulate the Blue Nile River.

The measured flow and developed suspended sediment concentration were used to simulate the morphodynamic in the Blue Nile River using the coupling model Sobek Rural DelWAQ. The characteristic size of the sediment in the bed at different points along the river is the necessary initial condition of the model (see Figure 2.25 in chapter 2). The results of 10 days average sediment concentration measured in 2009 at Kessie Bridge, Bure Bridge and El Deim along the Blue Nile River are shown in Figure 6.4.

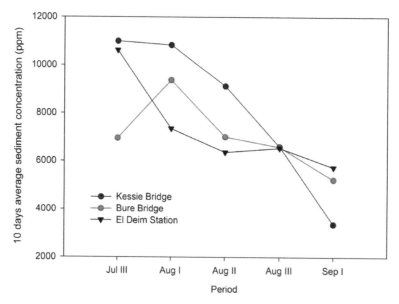

Figure 6.4: 10 days average sediment concentration along Blue Nile River at Kessie Bridge, Bure Bridge and El Deim Station (measured in 2009).

The characteristics of suspended sediment grain size distribution analysis along the river showed decreasing trend in the sand content towards the downstream direction and increasing trend for the silt and clay in the same direction. The results of the analysis of the Blue Nile River sieve analysis at Kessie Bridge, Bure Bridge, Chamoga Tributaries and El Deim station is presented in Figure 6.5.

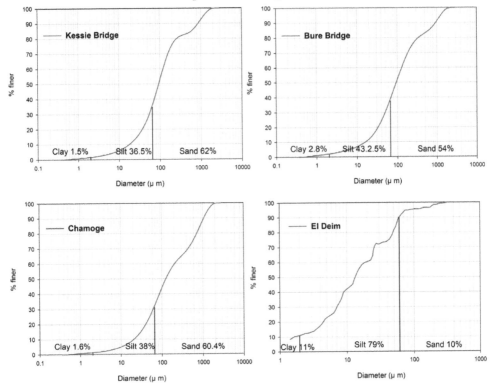

Figure 6.5: Suspended sediment at Kessie Bridge, Bure Bridge and El Deim Station; Grain size distribution.

6.5   MORPHODYNAMIC MODEL

6.5.1   Model setup

A 1D numerical model was constructed using the Sobek River/Morphology software in order to study sediment transport capacity and morphodynamic behavior of the Blue Nile River. The model was built to cover the Blue Nile River from Bahridar station, just downstream Lake Tana in Ethiopia (upstream boundary condition) to Roseires Dam in Sudan (downstream boundary condition) to reproduce the mechanisms of sediment transport and consequently the morphodynamic behavior of the river. The model geographic area was defined in Figure 6.6.

Four sediment fractions in the bed as separate substances SEDS01, SEDS02, SEDS03 and SEDS04 in the mixing layer or top-layer of the bed were defined. The volumes of these bed substances are affected by the sediment-transport capacity processes (bed-load or total-load transport formula) as well as interaction with the substances in the water column (Partheniades-Krone for clay, and Galappatti for sand).

Four sediment fractions in the water column as separate substances SEDW01, SEDW02, SEDW03 and SEDW04 (in suspension between bed and water surface) are defined corresponding to sediment fractions in bed. The concentrations of these substances are affected by advection-diffusion processes as well as interaction with the substances in the bed.

Based on the mass balances of the DelWAQ segment the processes affecting the volumes of each substance will lead to a bed-volume change of each fraction and a total volume change. For stability of the coupled flow/morphology simulations it is necessary that bed update by these bed-volume changes is done in a numerically stable and mass conserving way.

The morphological model was set using the available and newly collected sediment concentration and flow data which is further used to develop sediment discharge curve. The morphodynamic model was calibrated in the period 1993 -2005 and validated in the period 2005-2007 at Roseires Lake based on the available measurement and bathymetric surveys.

The hydraulic structure of Roseires and Renaissance Dams were simulated using a compound structure, consisting of a weir to control the water level in the lake and a one gate to control the release flow controlled. The temporal variant of the lake water level and the gate opening is simulated by time controller.
The crest widths of the weirs have been chosen in order to obtain a good profile of the water level upstream the dam; the Roseires weir is 3000 m wide and the Renaissance weir is 5000 m wide.

Upstream boundary condition of the hydrodynamic model is the daily flow time series released downstream Lake Tana and the downstream boundary condition is the daily water level measured at Roseires Reservoir. The discharge from the tributaries is used as lateral flow entering the system including Andassa, Mendel, Tigdor, Muga, Suha; etc.. in the North Gojam sub basin, Jemma, Welaka and Jemma in the left bank to the East, Jedeb, Chamoga, Temcha, Birr, Yeda among others in South Gojam sub basin, Muger, Guder , Fincha, Didessa, Anger, Dabus and Beles.
In the morphodynamic model, the continuity equation requires one boundary condition for each inflowing boundary, which means that at the upstream

boundaries of the model boundary conditions are imposed. Two parameters have to be specified for the boundary conditions; the bed level changes (dzBC) and the boundary segments (bound). The inflow boundary and outflow of the model were marked by setting the parameter "bound" to a value of 1 which means that this segment is boundary segment. The bed level change "dzBC" was specified by giving a value 0 for the parameter to fix the bed on the upper boundary.

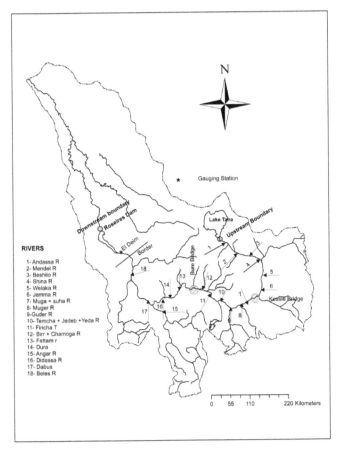

RIVERS

1- Andassa R
2- Mendel R
3- Beshilo R
4- Shina R
5- Welaka R
6- Jemma R
7- Muga + suha R
8- Muger R
9- Guder R
10- Temcha + Jedeb +Yeda R
11- Fincha T
12- Birr + Chamoga R
13- Fettam r
14- Dura
15- Angar R
16- Didessa R
17- Dabus
18- Beles R

Figure 6.6: Sobek morphology, the model geographic area.

For the suspended load sediment concentrations per fraction have to be specified at the inflow boundary. Time series sediment concentration in gm/m³ was used in the upstream boundary, lateral discharge and downstream boundary. Suspended sediment concentrations measured in gauging stations just during the wet season were used to generate rating curves followed by estimating the sediment concentration for measured discharges.

The selection of temporal and spatial parameter is important to obtain a good stability and accuracy of the model. The space step has to be chosen considering that a too large step leads to lack of the model accuracy. On the other hand, a too small step does not add useful information to obtain a good modelling and increasing the computational time. The space step of the spatial grid has been chosen to be 2000 m all over the river reach excluding the structures where grid space was used to be 0.5 m upstream and downstream the structure.

6.5.2   Model calibration

Some parameters are adjusted to obtain a good resemblance between model results and prototype values, with respect to physical accepted values. The hydrodynamic model was calibrated for the period 1990-1993. Several runs were carried out with different Manning coefficients, ranging between 0.02 s/m$^{1/3}$ and 0.035 s/m$^{1/3}$ with the aim to optimize the reproduction of the observed water levels and discharges at El Deim station. The adjusted values of Manning coefficients was found to be 0.03 s/m$^{1/3}$ in the main stream and 0.035 s/m$^{1/3}$ in the flood plain.

The results of hydrodynamic model calibration in terms of water levels and discharges are shown in Figures 6.7. The comparison between modeled and measured discharges and water levels showed a good agreement between model results and measurements.

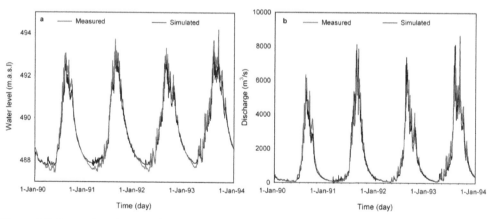

Figure 6.7: Comparison between the measured and simulated water levels (a) and discharges (b) at El Deim station during the period 1990-1993.

The morphodynamic calibration was based on the period 1992-2005 for which bathymetric surveys in Roseires Reservoir and daily measured sediment concentration in 1993 are available. The calibration parameters are grain size of the sediment along the river (D$_{50}$), packing factor, chosen 0.2 considering that the material is poorly sorted (large variance in sediment size) and sand bed dry density which is adapted to be 2350 kg/m^3, which is acceptable since the non-cohesive sediment in the reservoir is graded sand. The dry density determines the thickness of the total sediment layer by calculating the sediment mass and dividing it by the dry density.

The critical bed shear stress of sedimentation is kept at 1000 N/m^2 which means the suspended sediment is always allowed to deposit. On the other hand, the critical shear stress for erosion is increased to 1 N/m^2 which means unless the bed shear stress exceeds this value, no bed erosion takes place; in addition the erosion parameter rate is reduced up to 2 $mg/m^2/s$.

The total cumulative deposited volume of sediment computed from the model (1993-2005) is 234.5 million m^3, while the sediment volume measured in the bathymetric survey of 1993 and 2005 was found to be 168 million m^3. The result showed that the average sediment deposition rate in Roseires Reservoir during the period 1993 to 2005 is 19.4 million m^3/year, whereas the bathymetric surveys during the same period showed a rate of 14 million m^3/year.

Comparison between the measured and simulated suspended sediment concentrations at El Deim station is shown in Figure 6.8. It should be noted that the measurement were carried out during the wet season, however, we measured average suspended sediment concentration of 0,024 kg/m^3 during the low season

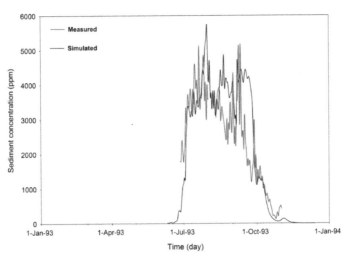

Figure 6.8: Comparison between the measured and simulated sediment concentration at El Deim station during 1992.

6.5.3   Model validation

The hydrodynamic model was validated for the period 2004-2007. The results of model validation in terms of water levels and discharges are shown in Figures 6.9. The comparison between modeled and measured discharges and water levels showed a good agreement between model results and measurements.

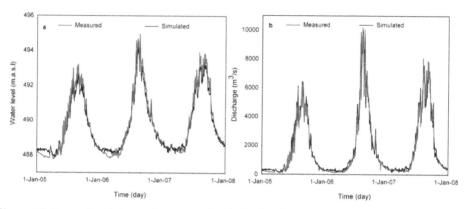

Figure 6.9: Comparison between the measured and simulated water levels (a) and discharges (b) at El Deim station during the period 2004-2007.

The total cumulative deposited volume of sediment computed from the model (2005-2007) is 27.4 million m³, while the sediment volume measured in the bathymetric survey of 2005 and 2007 was found to be 19 million m³. The result showed that the average sediment deposition rate in Roseires Reservoir during the period 2005 to 2007 has reduced to 9 million m³/year, whereas the bathymetric surveys during the same period showed a rate of 7 million m³/year. The total cumulative deposited volume of sediment inside Roseires Reservoir during the period 1993 to 2007 is depicted in Figure 6.10.

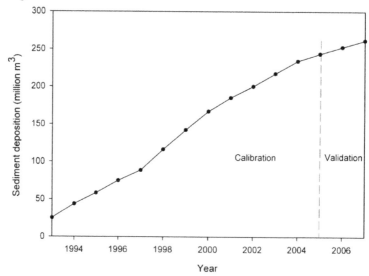

Figure 6.10: Cumulative deposited volume of sediment inside Roseires Reservoir during the period 1993 to 2007.

6.5.4   Impact of new developments

After the model was calibrated and validated, it was used to assess the morphological impacts of the planned developments, including Roseires Dam heightening and the construction of the Grand Ethiopian Renaissance Dam (GERD). Salini and SP (2010) studied the operation and filling of Grand Ethiopian Renaissance Dam.   According to their study, the impounding will start at the beginning of July 2013 and competing in 24th of February 2018 as described in detail in table 6.2.

Table 6.2: Grand Ethiopian Renaissance Dam- Impounding programme.

Impounding evaluations (m.a.s.l)	Starting day	Ending day
Natural to 515	1-July-2013	28-Jun-2014
515 to 540	30-Jun-2014	28-Jun-2015
540 to 590	27-Jun-2015	28-Jun-2016
590 to 640	29-Jun-2016	24-Feb-2018

The operation of GERD reservoir for average year was carried out based on average monthly inflows, evaporation losses and rainfall over the reservoir and reservoir starting and ending at maximum elevation. The summary of all the main parameters of the reservoir routing including reservoir elevation, energy production, turbined volumes and runoff are illustrated in Figure 6.11. In these calculations, the reservoir full storage level was assumed to be 640 m.a.s.l, the average operation level of 632 m.a.s.l and the minimum operation level of 622 m.a.s.l (Salini and SP, 2010).

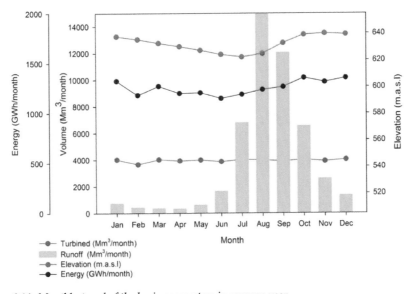

Figure 6.11: Monthly trend of the basic parameters in average year.

In order to study the effects of Renaissance Dam construction and Roseires Dam heightening, the model was run considering the new operation of Roseires (Figure 6.1), the filling (Table 6.2) and operation of Renaissance (Figure 6.11). The calibrated model Sobek morphology was set up using the newly collected cross sections measured in the framework of this study both in Ethiopia and Sudan. The measured cross section in Roseires Reservoir in Sudan covers the expected surface area of the reservoir at an elevation of 491 m.a.s.l. However, few cross sections in Ethiopia at Renaissance Dam are measured and some are obtained from Salini report (2010). The calibrated model was run using the same data from 1993 to 2007.

The result showed that the annual sediment deposition inside Renaissance Dam varies with time and has maximum and minimum values of 45 and 17 million m³/year in 1998 and 2003 respectively. The variation of sediment deposition inside Renaissance Dam is shown in Figure 6.12 a. The average sediment deposition rate in Renaissance construction during the period 1993 to 2007 was found to be 27 million m³/year. Some studies showed that this number is small

The average sediment deposition rate in Roseires Reservoir after heightening and Renaissance construction during the period 1993 to 2007 was found to be 2 million m³/year. Compression of sediment accumulated inside Roseires Dam before and after heightening and the Renaissance Dam is shown in Figure 6.12 b. The annual sedimentation rates at Roseires Reservoir has been reduced significantly.

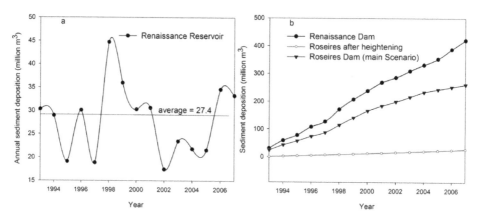

Figure 6.12: The variation of sediment deposition inside Renaissance Dam (a) and accumulation of sediment deposition inside Roseires Dam before and after heightening and Renaissance Dam (b).

Basin management intervention involves introducing the best erosion control practices to reduce soil erosion and sediment transport will be the best mitigation measure both for sedimentation problems in Sudan and erosion problem in Ethiopia. The model was applied to simulate the impacts of erosion control practices implementation to reduce the erosion in the most three sub basin that produce the sediment namely Jemma, Didessa and South Gojam. Five different scenarios were applies as follows:

- ✓ Scenario 0, the conditions after Roseires Dam heightening and Renaissance Dam construction is considered.
- ✓ In Scenario 1, it is assumed that no sediment entering the system from Jemma sub basin.
- ✓ In Scenario 2, it is assumed that no sediment entering the system from Didessa sub basin.
- ✓ In Scenario 3, it is assumed that no sediment entering the system from South Gojam sub basin.
- ✓ In Scenario 4, it is assumed that no sediment entering the system from above three sub basins.

The model results showed that the accumulative sediment deposited inside Renaissance Reservoir during 1993 to 2007 is 421.2 million m³. However, sediment deposition inside the reservoir has reduced by 17%, 20%, 6% and 43.4 % for scenarios1, 2, 3 and 4 respectively during the same period (Figure 13).

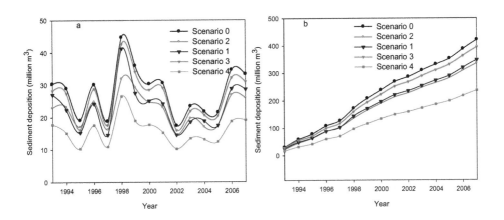

Figure 6.13: Comparison between sediment deposition inside Renaissance Dam for different erosion control practices scenarios (a) annual and (b) accumulation.

6.6 CONCLUDING REMARKS

The unavailability of good field data reflects on the model accuracy and output reliability. In this study, data were not available in sufficient detail, and were limited in terms of quality and extent. River cross sections, bed and bank soil materials collected during 2009 field campaign from scattered locations were used to set up the model through interpolation for the unmeasured reaches.

The model was calibrated during the period 1993 to 2005 did not resemble Roseires Reservoir sedimentation along its life time because there is no bathymetric data for the period 1966 when it is constructed. The sedimentation rate was reported for the first 10 years to be 55 million m^3/year.

In this study one dimensional model was used to study the sedimentation processes in the Blue Nile River including reservoirs. However, two dimensional morphodynamic model is needed to simulate the sedimentation processes in Grand Ethiopian Renaissance Dam.

Chapter 7
DISCUSSION AND CONCLUSIONS

In the Blue Nile Basin, erosion, sediment transport and sedimentation are causing considerable loss of fertile soil in the upper basin, loss of storage capacity in reservoirs and loss of conveyance in irrigation canals in the lower basin.

The analysis of data collected over the entire river basin at several locations along the main river and tributaries showed that sediment and flow distribution are highly variable with time, and that the highest sediment concentration occurs in July one month before the peak flow. This happens because most of the lands are rainfed cultivated and at the beginning of the rainy season, the soil is bare or with less vegetation cover that can easily be eroded.

This study has provided answers to the following general questions:

> **Does the erosion in the upper Blue Nile River catchment in Ethiopia result in increased sediment load and increased sedimentation in Sudan?**

The research work included the analysis of the mineral content of the sediments deposited in the Roseires Reservoir, in Sudan, and of the soil samples from 28 different locations in the upper Blue Nile basin, selected among the eroding areas, in Ethiopia (Chapter 5). The results show that most sediment deposited inside Roseires Reservoir originates from the sub-basins of Jemma, Didessa and South Gojam.

The analysis of historical land use changes in these sub-basins, based on satellite images taken in 1973, 2000 and 2010 (Sub-chapter 3.4), shows significant changes from natural forests to cultivated soils.

The sediment balances derived for each sub-basin of the Blue Nile basin from the analysis of old and newly acquired field data and numerical modelling (Chapter 3), show that the quantitative contribution of these sub-basins to the amount of sediment reaching Sudan is 37% of the total, the latter being 130-160 million tonne/year.

The conclusion is that sedimentation in Sudan is for a large part caused by soil erosion in the upper catchment.

> **What are the hydromorphological and sedimentological characteristics of the Blue Nile River system in Ethiopia and Sudan?**

The study presents a basin-scale description of the hydromorphological and sedimentological characteristics of the Blue Nile River network, derived from the analysis of field data and numerical modelling: the flow and sediment balances are given in Chapter 3; the hydrodynamic characteristics and the cross-sectional profiles of the Blue Nile River network in Chapter 4; the sediment fingerprints in Chapter 5.

In general, the study was characterized by lack of existing data and by the difficulty of collecting new ones. Yet, the outcomes provide a comprehensive picture of the hydromorphological and sedimentological characteristics of the Blue Nile River network, which could be used as basis for more detailed works in the future.

This result was made possible by the comprehensive approach used, including the analysis of existing historical data and satellite images, the acquisition and analysis of new field data, and the use of different numerical tools (SWAT, SOBEK, Delft3D Delwaq) selected based on the scale of the investigation and the simulated processes.

> **What will be the impact of the Grand Ethiopian Renaissance Dam (GERD) and heightening of Roseires Dam on sediment processes?**

The construction of Grand Ethiopian Renaissance Dam (GERD) in Ethiopia will have significant impacts on downstream sediment transport rates, which will drastically decrease. An important positive effect will be the reduction of sedimentation rates inside Roseires Reservoir.

This work also answers the following specific questions:

> **How can we determine sediment origin and timing of sediment transfer along the Blue Nile River?**

Sedimentation processes inside Roseires Reservoir were studied over the period 1985 - 2007 to analyse the characteristics of the deposited sediment and to estimate the sedimentation rates within the reservoir. Samples from the cores were analyzed in combination with the upper basin soils to identify the source. The analysis was conducted combining computer modelling (2D modelling using state of the art software, Delft3D) and field survey of both sediment quality and quantity. The model allowed studying the contribution of two sediment types, sand and fines, both transported by the Blue Nile into the reservoir. Sand is deposited especially during high flows; fines during low flow periods. This creates soil stratification inside the reservoir allowing the recognition of specific wet or dry years.

The model showed variable sedimentation rates during the 22 years between 1985 and 2007. In particular, in 1988 the reservoir gained more storage capacity. This can be referred to the extremely high flood of 1988, which caused net soil erosion, especially in the area just upstream of the dam. The computed sedimentation rates have the same order of magnitude as the measured ones for the period 1985-1992, but they are approximately twice as large for the period 1992-2007. Using the same setting of parameters for the following fifteen years resulted in model overestimation of the sedimentation rates. This could be caused by the overestimation of the sediment inputs. For this reason suspended sediment should be carefully measured in the future to minimize uncertainties related to sediment input for future estimations. Only one measurement of suspended solids was carried out, in 2011, during the low flow season, whereas all other measurements had been carried out during the high flow season. More measurements are required during the low water season, at least for modelling purposes.

Reservoir soil stratification was studied in the areas that were always subjected to deposition. In Roseires Reservoir, a main natural channel having a sinuous shape meanders through the deposited sediment. This channel was found to shift laterally at certain locations. The low model resolution, caused by the relatively large model grid size, made it difficult to correctly detect the channel movement inside the reservoir. So, to be on the safe side, coring was carried out far from the main channel. Keeping this in mind, the model allowed identifying two promising coring areas that were neither subject to net erosion nor to bar migration during the life span of the reservoir.

The mineral content of the soils and rock both in upper basin (source) and sediment deposited in Roseires Reservoir (sink) was used to analyze erosion/sedimentation in the Blue Nile River Basin. A X-Ray Diffraction (XRD) tool was used to determine the mineral content in 63 samples from the whole basin. Cluster analysis was applied to identify clusters of locations by the development of an algorithm to define a group with similar mineral characteristics. The technique was applied to compare the six minerals content resulted from XRD analysis to define similarity between two data points. First; cluster analysis was performed for the whole samples to quantify how "similar" two data points to know the overall pattern of mineral contents. The results showed that most of sediment entering Sudan is similar to that from the Jemma, Didessa and South Gojam. Second the cluster analysis was performed for the layers of sediment deposited in Roseires Reservoir every 1 m and samples from the upper basin. In most cases, the results suggest that most of the sediment deposited inside Roseires Reservoir is originated by erosion from Jemma, Didessa and South Gojam sub basins in the upper Blue Nile basin and these have similar characteristics. The cluster analysis proved to be a useful semi-quantitative technique for analyzing the data and determining linkages between the sediment deposited inside Roseires Reservoir and its source in the Upper Blue Nile Basin.

The outputs of this study highlight the most critical sub basins for sediment input into the Blue Nile system. This becomes even more relevant in the construction of the Grand Ethiopian Renaissance Dam. When the construction of this dam is finished, it will be the largest dam in Africa and it will restrict most of sediment from flowing downstream. The sedimentation issue may be a major problem for adequate operation of the dam.

The results of the study do not allow to directly derive the timing of the sediment transfer process from the upper basin to Roseires Reservoir. This needs further research, which should be based on extensive coring of the reservoir soil detecting specific years from the deposit layers and the study of the soil erosion process in the upper basin. Modelling of sediment transfer becomes an option only when enough data are available.

How can we estimate the water and sediment balances along the Blue Nile River Basin both in Ethiopia and Sudan?

For the first time, this study provides the sediment balance of the entire Blue Nile Basin. A field campaign was conducted to collect sediment concentration data from several locations. In addition, the Soil and Water Assessment Tool (SWAT model) was used to generate runoff estimates for the un-gauged basins.

In addition, using the bathymetric survey of Roseires, it was possible to derive the sedimentation rates inside the reservoir to show the extent of sediment yield and the massive sediment volume at the outlet of Blue Nile from the Ethiopian highland. It was found that the three sub basins that exported the highest sediment yields are Jemma sub basin,19-27 million tonne/year (1400-2000 tonne/km^2/year), South Gojam sub basin, 11-17 million tonne/year (680-1000 tonne/km^2/year) and Didessa sub basin, 12-14 million tonne/year (592-710 tonne/km^2/year). In contrast, the Anger sub basin generated a relative low sediment export, 2-3 million tonne/year (200-330 tonne/km^2/year).The general trends observed for land-cover were obtained by analyzing the 1973, 2000 and 2010 satellite images of the Jemma, Didessa and South Gojam sub basins. This shows a decrease in wooded grassland, woodland, shrubs and bushes and a matching increase in rainfed cropland. This tendency may explain the land degradation and hence the sediment produced by these sub basins.

What types of methods and tools can be used to follow the sediment to the sinks?

One-dimensional, 1D, models are more suitable for long term simulations of channel cross-sectional change of a long study reach than two or three dimensional models because of computational time and because they need less input data.

A 1D model, Sobek River/Morphology was used in this thesis to study the sediment transport rates in the entire Blue Nile River reach upstream Roseires Dam (more than 1000 km) over several years to predict channel geometry in a semi-two dimensional. However, like all 1D numerical models, many sediment processes are ignored because the model does not directly simulate transverse sediment movement and calculates a single average shear stress for a cross section. In Roseires Reservoir, the sedimentation trends reproduced by the model showed a general resemblance with the field data and with the typical patterns resulting from the sedimentation process in reservoirs.

The model used is not capable of following the path of the sediment particles and estimate their transfer time from source to sink. For this other modeling techniques are necessary.

How can we develop a basin scale model for Blue Nile capable of sediment transfer?

The model was built to cover the Blue Nile River from Bahridar station, just downstream Lake Tana in Ethiopia (upstream boundary condition) to Roseires Dam in Sudan (downstream boundary condition) to reproduce the mechanisms of sediment transport and consequently the morphodynamic behaviour of the river.

The hydrodynamic model was built on cross-sectional data from the 1990s for the Sudanese part of the river and from 2009 for the Ethiopian part (data collected in the framework of this study). Calibration was performed on the period 1990 to1993 and validation on the period 1994 to 1996. The simulations showed good correlation coefficients between computed and measured water levels and discharges for both the calibration and validation runs. However, since most data used for model set up, calibration and validation are from the 1990's, the model strictly represents the Blue Nile river behaviour 15-20 years ago. To check its performance on recent years, the model was then applied in the recent years 2008, 2009 and 2010. The comparison between computed and measured water levels and discharges shows that it performs well also for the present situation. For this, the morphological changes occurred in the last 20 years do not seem to have impact on the hydrodynamic behaviour of the Blue Nile River system in a relevant way.

The calibrated hydrodynamic model was coupled with morphological model Delwaq using the available and newly collected sediment concentration and flow data which is further used to develop sediment discharge curve. The morphodynamic model was calibrated in the period 1993 -2005 and validated in the period 2005-2007 at Roseires Lake based on the available measurement and bathymetric surveys.

How can we assess the new developments the Blue Nile River system?

The calibrated Sobek River/Morphology model was used to assess the new development along the river system such as Roseires Dam heightening and Grand Ethiopian Renaissance Dam after construction. The construction of Grand Ethiopian Renaissance Dam (GERD) in Ethiopia will have significant impacts on the downstream sediment transport. Also the model was used to assess the impact of different erosion control practices on the sedimentation rates inside these reservoirs.

The model showed that the average sediment deposition rate in Roseires Reservoir after heightening and Renaissance construction during the period 1993 to 2007 was found to be 2 million m^3/year. The bathymetric surveys of 2005 and 2007 showed the average sediment deposition rate in Roseires Reservoir is about 6.8 million m^3/year. The annual sedimentation rates at Roseires Reservoir have been reduced significantly.

Basin management intervention involves introducing the best erosion management practices to reduce soil erosion and sediment transport will be the best mitigation measure both for sedimentation problems in Sudan and erosion problem in Ethiopia. The model was applied to simulate the impact if erosion management practices are implemented in the most three sub basins that produce the sediment namely Jemma, Didessa and South Gojam. The model results showed that the reduction of accumulative sediment deposited inside Renaissance Reservoir (GERD) varied (6% to 43.4 %).

Summary of conclusions

- ✓ The sediment and flow distribution along Blue Nile River and its tributaries are highly variable with time, and the highest sediment concentration occurs in July one month before the peak flow.
- ✓ For the first time, this study provides sediment balance of the entire Blue Nile Basin via integration of available and newly collected data and Soil and Water Assessment Tool (SWAT) model.
- ✓ This sediment balance study showed that the three sub basins exported the highest sediment yields are Jemma, South Gojam and Didessa sub basins, whereas, Anger sub basin generated a relative low sediment export.

✓ The general trend of land-cover changes for Jemma, Didessa and South Gojam sub basins was obtained by analyzing the satellite imageries 1973, 2000 and 2010. The results showed decreasing trend in wooded grassland, woodland, shrubs and bushes and a matching increasing in rainfed cropland.

✓ The hydrodynamic characteristics were studied by means of a model covering the entire Blue Nile River system. The hydrodynamic model Sobek Rural was used to quantify the availability of the water resource throughout the year.

✓ The sedimentation processes inside Roseires Reservoir was simulated using Delft3D model during the 22 years between 1985 and 2007. Reservoir soil stratification was studied in the areas that were always subjected to deposition. The model allowed identifying two promising coring areas that were neither subject to net erosion nor to bar migration during the life span of the reservoir.

✓ The mineral analysis for the soils and rock both in upper basin (source) and sediment deposited in Roseires Reservoir (sink) was performed using X-Ray Diffraction (XRD) tool. Cluster analysis was applied to identify clusters of locations by the development of an algorithm to define a group with similar mineral characteristics.

✓ The cluster analysis results showed that most of sediment deposited in Roseires Reservoir is similar to that from Jemma, Didessa and South Gojam.

✓ Sobek morphology model showed the annual sedimentation rates at Roseires Reservoir have been reduced significantly after Roseires Heightening and GERD construction.

✓ The model results showed that the accumulative sediment deposited inside Renaissance Reservoir during 1993 to 2007 is 421.2 million m^3. The model was applied to simulate the impact if erosion control practices in the most three sub basin that produces the sediment namely Jemma, Didessa and South Gojam. The resulted showed that sediment deposition inside the reservoir has reduced by 17% (if no sediment from Jemma), 20% (if no sediment from Didessa), 6% (if no sediment from South Gojam) and 43.4 % (if no sediment from Jemma, Didessa and South Gojam).

REFERENCES

Abbaspour, K.C., Faramarzi, M., Ghasemi, S.S. and Yang, H., 2009. Assessing the impact of climate change on water resources in Iran. Water Resources Research, 45(10).

Abbaspour, K.C., Johnson, A. and van Genuchten, M.T., 2004. Estimating uncertain flow and transport parameters using a sequential uncertainty fitting procedure. Vadose Zone J, 3: 1340-1352.

Abbaspour, K.C., Yang, J., Maximov, I., Siber, R., Bogner, K., Mieleitner, J., Zobrist, J. and Srinivasan, R., 2007. Modelling hydrology and water quality in the pre-alpine/alpine Thur watershed using SWAT. Journal of Hydrology, 333(2-4): 413-430.

Abd Alla, M.B. and Elnoor, K., 2007. Hydrographic Survey of Roseires Reservoir, Ministry of Irrigation and Water Resources,Khartoum, Sudan, Khartoum.

Abdalla, S.H., 2006. Socio-economic and environmental impact of sedimentation in Sudan., International Sediment Initiative Conference. UNESCO Chair in Water Resources., Khartoum, Sudan, pp. 480-491.

Abdelsalam, A.A. and Ismail, U.H., 2008. Sediment in the Nile River system. Consultancy Study requested by UNESCO.

Abdo, K.S., Fiseha, B.M., Rientjes, T.H.M., Gieske, A.S.M. and Haile, A.T., 2009. Assessment of climate change impacts on the hydrology of Gilgel Abay catchment in Lake Tana basin, Ethiopia. Hydrological Processes, 23(26): 3661-3669.

Ahmed, A.A., Ibrahim, S.A.S., Ogembo, O., Ibrahim, A.A. and Crosato, A., 2010. Nile River bank erosion and protection, Hydraulic Research Institute, Cairo.

Ahmed, A.A. and Ismail, U.H.A.E., 2008. Sediment in the Nile River System., UNESCO and International Sediment Initiative (ISI), Khartoum.

Al-Jaroudi, S.S., Ul-Hamid, A., Mohammed, A.I. and Saner, S., 2007. Use of X-ray powder diffraction for quantitative analysis of carbonate rock reservoir samples. Powder Technology, 175(3): 115-121.

Aldenderfer, M. and Blashfield, R., 1984. Cluster Analysis. SAGE Publication Inc., Newbury Park,CA, 44 pp.

Ali, Y.S.A. and Crosato, A., 2013. Sediment balances in the Blue Nile River Basin. In: A. Crosato (Ed.), Netherlands Centre for River Studies, UNESCO-IHE, Delft, the Netherlands.

Ali, Y.S.A., Crosato, A., Mohamed, Y.A., Abdalla, S.H., Wright, N.G. and Roelvink, J.A., 2013a. Morphodynamics impacts of the Grand Renaissance Dam construction and Roseires Dam heightening along Blue Nile River, The New Nile Perspectives conference. Hydraulics Research Center and UNESCO-IHE, Khartoum, Sudan.

Ali, Y.S.A., Omer, A.Y.A. and Crosato, A., 2013b. Modelling of sedimentation processes inside Roseires Reservoir (Sudan), The 8th Symposium on River, Coastal and Estuarine Morphodynamics (RCEM 2013), University of Cantabria ,Santander, Spain.

Allan, J., 2009. Nile Basin Asymmetries: A Closed Fresh Water Resource, Soil Water Potential, the Political Economy and Nile Transboundary Hydropolitics. In: H. Dumont (Ed.), The Nile. Monographiae Biologicae. Springer Netherlands, pp. 749-770.

Allen, P.A., 1997. Earth Surface Processes. John Wiley & Sons, Hoboken, NJ.

Anderson, G.M. and Burnham, C.W., 1965. The solubility of quartz in super-critical water. American Journal of Science, 263: 494-511.

Annandale, G.W., 1987. Reseroir Sedimentation. Developments in water science 29, 29. Elsevier Science Publishers B. V, New York.

Ariathurai, R. and Arulanandan, K., 1978. Erosion rates of cohesive soils. Journal of the Hydraulics Division, 104(2): 279-283.

Army, U.S., 1993. The Hydraulic Engineering Center, *HEC-6 Scour and Deposition in Rivers and Reservoirs*. In: M.r. *Usev's Manual* (Ed.), *Usev's Manual*, March 1977 (revised 1993), USA.

Arnold, J.G., Abbaspour, K.C., Srinivasan, R., Santhi, C. and Kannan, N., 2012. SWAT: Model use, calibration, and validation. Transactions of the ASABE, 55(4): 1491-1508.

Arnold, J.G. and Fohrer, N., 2005. SWAT2000: current capabilities and research opportunities in applied watershed modelling. Hydrological Processes, 19(3): 563-572.

Arnold, J.G., Srinivasan, R., Muttiah, R.S. and Williams, J.R., 1998. Large area hydrologic modeling and assessment part I: Model development. Journal of the American Water Resources Association, 34(1): 73-89.

Ascaso, F.J. and Blasco, J., 2012. X-ray Diffraction Analysis to Clarify the Unusual Origin of an Intraocular Foreign Body. British Journal of Medicine & Medical Research, 2(2): 228-234.

Asselman, N.E.M., 2000. Fitting and interpretation of sediment rating curves. Journal of Hydrology, 234(3–4): 228-248.

Awulachew, S.B., Erkossa, T., Smakhtin, V. and Fernando, A., 2010. Improved water and land management in the Ethiopian highlands: its impact on downstream stakeholders dependent on the Blue Nile. International Water Management Institute.

Awulachew, S.B., McCartney, M., Steenhuis, T.S. and Ahmed, A.A., 2008. A review of hydrology, sediment and water resource use in the Blue Nile Basin. International Water Management Institute, Addis Ababa, Ethiopia.

Ayalew, L. and Yamagishi, H., 2004. Slope failures in the Blue Nile basin, as seen from landscape evolution perspective. Geomorphology, 57(1–2): 95-116.

Bagnold, R., 1966. An approach to the sediment transport problem from general physics: US Geol. Survey Prof. Paper, 422(1): 37.

Balthazar, V., Vanacker, V., Girma, A., Poesen, J. and Golla, S., 2013. Human impact on sediment fluxes within the Blue Nile and Atbara River basins. Geomorphology, 180–181: 231-241.

Bashar, K.E., ElTahir, E.T.O., Fattah, S.A., Siyam, A.M. and Crosato, A., 2010. Nile Basin Reservoir Sedimentation Prediction and Mitigation. , Hydraulic Research Institute, Nile Basin Building Capacity Network, Cairo.

Bashar, K.E. and Eltayeb, A., 2010. Sediment Accumulation in Roseires Reservoir. Nile Basin Water Science and Engineering Journal, 3(3): 46-55.

BBC, 2010. East Africa seeks more Nile water from Egypt, 14 May 2010 http://news.bbc.co.uk/2/hi/africa/8682387.stm.

BCEOM, 1998. Abay River Basin Master Plan Project - Phase 3. Ministry of Water Resources, Addis Ababa, Ethiopia.

BCEOM, 1999. Abbay River Basin Master Plan Project - Phase 3, Ministry of Water Resources, Addis Ababa, Ethiopia.

Best, J., 2005. The fluid dynamics of river dunes: A review and some future research directions. Journal of Geophysical Research, 110(F04S02): 1-21.

Betrie, G.D., Mohamed, Y.A., Van Griensven, A. and Srinivasan, R., 2011. Sediment management modelling in the Blue Nile Basin using SWAT model. Hydrology and Earth System Sciences, 15(3): 807-818.

Bewket, W., 2002. Land Cover Dynamics Since the 1950s in Chemoga Watershed, Blue Nile Basin, Ethiopia. Mountain Research and Development, 22(3): 263-269.

Bhutiyani, M.R., 2000. Sediment load characteristics of a proglacial stream of Siachen Glacier and the erosion rate in Nubra valley in the Karakoram Himalayas, India. Journal of Hydrology, 227(1–4): 84-92.

Billi, P. and el Badri Ali, O., 2010. Sediment transport of the Blue Nile at Khartoum. Quaternary International, 226 (2010), 12-22.

Blanckaert, K., Glasson, L., Jagers, H.R.A. and Sloff, C.J., 2003. Quasi-3D simulation of flow in sharp open-channel bends with equilibrium bed topography, in RIver, Coastal and Estuarine Morphodynamics: RCEM 2003, 1-5 Sept. 2003, Barcelona, Spain, eds. A.Sanchez-Arcilla & Bateman A., IAHR, Vol. I, PP. 652-663.

Block, P.J., 2007. Integrated management of the Blue Nile Basin In Ethiopia: Hydropower and irrigation modelling. International Food Policy Research Institute, Washington, DC, pp. 1-21.

Blom, A., 2008. Different approaches to handling vertical and streamwise sorting in modeling river morphodynamics. Water Resources Research, 44(W03415): 1-16.

Borland, W.M. and Miller, C.R., 1958. Distribution of sediment in large reservoirs. ASCE, Journal of the Hydraulics Division, 84(2): 1-18.

Brandt, S.A., 2000. Classification of geomorphological effects downstream of dams. CATENA, 40(4): 375-401.

Brown, C.B., 1943. Discussion of "Sedimentation in reservoirs, by J. Witzig", Proceeding of the American Society of Civil Engineers, pp. 1493-1500.

Brune, G.M., 1953. Trap efficiency of reservoirs. Trans. AGU, 34(3): 407-418.

Chu, T.W. and Shirmohammadi, A., 2004. Evaluation of the SWAT model's hydrology component in the piedmont physiographic region of Maryland. Trans. ASAE 47(4): 1057-1073.

Church, M.A. and Rood, K.M., 1983. Catalogue of Alluvial River Channel Regime Data. Department of Geography, University of British Columbia (report).

Churchill, M.A., 1948. DIsscussion of "Analysis and use of reservoir sedimentation data" by L. C. Gottschalk,, Federal Inter-Agency Sedimentation Conference, Washington, D. C, pp. 139-140.

Collins, A.L. and Walling, D.E., 2002. Selecting fingerprint properties for discriminating potential suspended sediment sources in river basins. Journal of Hydrology, 261(1–4): 218-244.

Collins, A.L., Walling, D.E., Webb, L. and King, P., 2010. Apportioning catchment scale sediment sources using a modified composite fingerprinting technique incorporating property weightings and prior information. Geoderma, 155(3–4): 249-261.

Conway, D., 1997. A water balance model of the Upper Blue Nile in Ethiopia. *Hydrological Sciences Journal*, 42(2): 265-286.

Conway, D., 2000. The Climate and Hydrology of the Upper Blue Nile River. Geographical Journal, 166(1): 49-62.

Conway, D., 2005. From Headwater Tributaries to Internstionsl River : Observing and Adapting to Climate ChangeVaribility and Change in Nile Basin. Global Environmental Change, 15: 99-114.

Conway, D. and Hulme, M., 1996. The impacts of climate variability and climate change in the Nile Basin on future water resources in Egypt. Water Resources Development 12(3): 277-296.

Cortés, J.A., Palma, J.L. and Wilson, M., 2007. Deciphering magma mixing: The application of cluster analysis to the mineral chemistry of crystal populations. Journal of Volcanology and Geothermal Research, 165(3–4): 163-188.

Crawford, C.G., 1991. Estimation of suspended-sediment rating curves and mean suspended-sediment loads. Journal of Hydrology, 129(1–4): 331-348.

Crowder, D.W., Demissie, M. and Markus, M., 2007. The accuracy of sediment loads when log-transformation produces nonlinear sediment load–discharge relationships. Journal of Hydrology, 336(3–4): 250-268.

Cullity, B.D. and Stock, S., 1956. Elements of X-ray Diffraction. Addison-Wesley Metallurgy. Addison-Wesley Publishing Company, Inc., Notre Dame, Indiana.

Dai, Z. and Liu, J.T., 2013. Impacts of large dams on downstream fluvial sedimentation: An example of the Three Gorges Dam (TGD) on the Changjiang (Yangtze River). Journal of Hydrology, 480(0): 10-18.

de Meijer, R.J., James, I.R., Jennings, P.J. and Koeyers, J.E., 2001. Cluster analysis of radionuclide concentrations in beach sand. Applied Radiation and Isotopes, 54(3): 535-542.

de Vente, J., Poesen, J. and Verstraeten, G., 2005. The application of semi-quantitative methods and reservoir sedimentation rates for the prediction of basin sediment yield in Spain. Journal of Hydrology, 305(1-4): 63-86.

de Vente, J., Verduyn, R., Verstraeten, G., Vanmaercke, M. and Poesen, J., 2011. Factors controlling sediment yield at the catchment scale in NW Mediterranean geoecosystems. Journal of Soils and Sediments, 11(4): 690-707.

Delft-Hydraulics, 1992. Sudan Flood Early Warning System, Technical report prepared for the Ministry of Irrigation and Water Resources. Sudan, Khartoum.

Demissie, M., Xia, R., Keefer, L. and Bhowmik, N., 2004. The sediment budget of the Illinois River, Illinois State Water Survey- 2204 Griffith Drive- Champaign, IL 61820-7495

Easton, Z.M., Fuka, D.R., White, E.D., Ahmed, A.A. and Steenhuis, T.S., 2010. A multi basin SWAT model analysis of runoff and sedimentation in the Blue Nile, Ethiopia. Hydrology and Earth System Sciences, 14(10): 1827-1841.

Easton, Z.M., Walter, M.T., Fuka, D.R., White, E.D. and Steenhuis, T.S., 2011. A simple concept for calibrating runoff thresholds in quasi-distributed variable source area watershed models. Hydrological Processes, 25(20): 3131-3143.

El Monshid, B.F., El Awad, O.M. and Ahmed, S.E., 1997. Environmental effect of the Blue Nile sediment on reservoirs and irrigation canals. 5th Nile 2002 Conference. Addis Ababa.

Elagib, N.A. and Mansell, M.G., 2000. Recent trends and anomalies in mean seasonal and annual temperatures over Sudan. Journal of Arid Environments, 45(3): 263-288.

Engelund, F. and Hansen, E., 1967. A Monograph on Sediment Transport in Alluvial Streams, Technisk Vorlaq, Copenhagen, Denmark.

ENTRO, 2007. Eastern Nile Watershed Management Project. Cooperative Regional Assessment (CRA) for Watershed Management. Distributive Analysis. , Eastern Nile Technical Regional Office (ENTRO). Addis Ababa, Ethiopia 2007.

Everitt, B., Landau, S. and Leese, M., 2001. Cluster Analysis., New York, Oxford University Press, USA.

FAO, 1995. Digital Soil Map of the World and Derived Soil Properties, Food and Agriculture Organization of the United Nations, Rome, Italy.

Ferguson, R.I., 1986. River Loads Underestimated by Rating Curves. Water Resour. Res., 22(1): 74-76.

Fournier, R.O. and Potter Ii, R.W., 1982. An equation correlating the solubility of quartz in water from 25° to 900°C at pressures up to 10,000 bars. Geochimica et Cosmochimica Acta, 46(10): 1969-1973.

Gamachu, D., 1977. Aspects of Climate and Water Budget in Ethiopia. Addis Ababa University, Addis Ababa.

Garzanti, E., Ando, S., Vezzoli, G., Ali AbdEl Megid, A. and El Kammar, A., 2006. Petrology of Nile River sands (Ethiopia and Sudan): Sediment budgets and erosion patterns. *Earth and Planetary Science Letters*, 252: 327-341.

Gebremicael, T.G., Mohamed, Y.A., Betrie, G.D., van der Zaag, P. and Teferi, E., 2013. Trend analysis of runoff and sediment fluxes in the Upper Blue Nile basin: A combined analysis of statistical tests, physically-based models and landuse maps. Journal of Hydrology, 482(0): 57-68.

Gismalla, Y.A., 1993. Bathymwtric Survey of Roseires Reservoir, Ministry of Irrigation and Water Resources, Wad-Madani (Sudan).

Gismalla, Y.A., 2009. Sedimentation Problems In the Blue Nile Reservoirs and Gezira Scheme: A Review. Gezira Journal of Engineering and Applied Science, 4(2): 1-19.

Goldich, S.S., 1938. A study in rock-weathering:. Journal of Geology, 46(1): 17-58.

Gottschalk, L., 1964. Reservoir sedimentation. Handbook of Applied Hydrology. New York: McGraw-Hill. Section, 17.

Goudie, A.S., 2005. The drainage of Africa since the Cretaceous. Geomorphology, 67(3–4): 437-456.

Green, J. and El-Moghraby, A., 2009. Swamps of the upper White Nile. In: H.J. Dumont (Ed.), The Nile Origin, Environments, Limnology and Human Use. Monographiae Biologicae, pp. 193-204.

Gupta, H., Kao, S.-J. and Dai, M., 2012. The role of mega dams in reducing sediment fluxes: A case study of large Asian rivers. Journal of Hydrology, 464–465(0): 447-458.

Hauschild, A. and Sloff, K., 2009. User-Manual Morphology in SOBEK-River, Deltares, Delft, the Netherlands.

Hefferan, K. and O'Brien, J., 2010. Earth Materials. A John Wiley & Sons, Ltd., Publication, UK.

Hefferan, K. and O'Brien, J., 2010. Earth Materials. A John Wiley & Sons, Ltd., Publication, UK.

Hirano, M., 1971. River bed degradation with armouring. Transactions of the Japan Society of Civil Engineers(3): 194-195.

Hu, C., Guo, Q., Chen, J. and Cao, W., 2010. Applications of numerical simulation to the sedimentation in the Sanmenxia reservoir and the Lower Yellow River. International Journal of Environment and Pollution, 42(1-3): 148-165.

Hussein, A.S., Bashar, K.E., Fattah, S.A. and Siyam, A.M., 2005. "Reservoir Sedimentation", Nile Basin Capacity Building Network (NBCBN), Khartoum, Sudan.

Hussein, A.S. and Yousif, D.M., 1994. Prediction of settling basins performance for very fine sediment, International conference on Efficient Utilization and Management of Water Resources in Africa, Khartoum, Sudan, pp. 75-85.

Ikeda, S., 1982. Lateral bed load transport on side slopes. Journal of the Hydraulics Division, 108 (11)(11): 1369-1373.

Jenkins, R., 2000. X-ray Techniques: Overview. In: R.A. Meyers (Ed.), Encyclopedia of Analytical Chemistry. John Wiley & Sons Ltd, Chichester, pp. 13269–13288.

Jensen, J.R., 2005. Introductory Digital Image Processing: A Remote Sensing Perspective. Prentice Hall, Upper Saddle River, New York.

Johnson, S.C., 1967. Hierarchical clustering schemes. Psychometrika, 32(3): 241-254.

Johnston, R. and McCartney, M., 2010. Inventory of Water Storage Types in the Blue Nile and Volta River Basins, IWMI Working Paper 140. International Water Management Institute, Colombo, Sri Lanka:.

Julien, P., Klaassen, G., Ten Brinke, W. and Wilbers, A., 2002. Case Study: Bed Resistance of Rhine River during 1998 Flood. Journal of Hydraulic Engineering, 128(12): 1042-1050.

Kanungo, T., Mount, D.M., Netanyahu, N.S., Piatko, C.D., Silverman, R. and Wu, A.Y., 2002. An efficient k-means clustering algorithm: Analysis and implementation. Pattern Analysis and Machine Intelligence, IEEE Transactions on, 24(7): 881-892.

Kaufman, L. and Rousseeuw, P.J., 2009. Finding groups in data: an introduction to cluster analysis, 344. Wiley. com.

Kettner, A., Restrepo, J. and Syvitski, J., 2010. A spatial simulation experiment to replicate fluvial sediment fluxes within the Magdalena River Basin, Colombia. The Journal of Geology, 118(4): 363-379.

Kim, U. and Kaluarachchi, J., 2008. Application of parameter estimation and regionalization methodologies to ungauged basins of the Upper Blue Nile River Basin, Ethiopia. Journal of Hydrology(362): 29-56.

Krone, R.B., 1962. Flume studies of the transport of sediment in estuarial shoaling processes, University of California, Hydraulic and sanitary engineering laboratory, Berkeley

Lesser, G.R., Roelvink, J.A., van Kester, J.A.T.M. and Stelling, G.S., 2004. Development and validation of a three-dimensional morphological model. Coastal Engineering, 51(8-9): 883-915.

Li, Q., Yu, M., Lu, G., Bai, X. and Xia, Z., 2011. Impacts of the Gezhouba and Three Gorges reservoirs on the sediment regime in the Yangtze River, China. Journal of Hydrology, 403(3-4): 224-233.

Magilligan, F.J. and Nislow, K.H., 2005. Changes in hydrologic regime by dams. Geomorphology, 71(1-2): 61-78.

Mamuse, A., Porwal, A., Kreuzer, O. and Beresford, S., 2009. A new method for spatial centrographic analysis of mineral deposit clusters. Ore Geology Reviews, 36(4): 293-305.

McCartney, M.P., Shiferaw, A. and Seleshi, Y., 2009. Estimating environmental flow requirements downstream of the Chara Chara weir on the Blue Nile River. Hydrological Processes, 23(26): 3751-3758.

Mehta, A.J., Hayter, E.J., Parker, W.R., Krone, R.B. and Teeter, A.M., 1989. Cohesive Sediment Transport. *Journal of Hydraulic Engineering*, 115(8): 1076-1093.

Meyer, P.E. and Müller, R., 1947. Formulas for bed load transport, 2nd IAHR congress, Stockholm, Sweden, pp. 39-64.

Middelkoop, H., 2002. Reconstructing floodplain sedimentation rates from heavy metal profiles by inverse modelling. Hydrological Processes, 16(1): 47-64.

Middelkoop, H. and Asselman, N.E.M., 1998. Spatial variability of floodplain sedimentation at the event scale in the Rhine–Meuse delta, The Netherlands. Earth Surface Processes and Landforms, 23(6): 561-573.

Milliman, J.D. and Farnsworth, K.L., 2011. River Discharge to the Coastal Ocean: A Global Synthesis Cambridge University Press., pp. 392.

Minitab, 1998. MINITAB Release 14, State College,PA. USA.

Montes, A.A., Crosato, A. and Middelkoop, H., 2010. Reconstructing the early 19th century Waal River by means of a 2D physics-based numerical model. Hydrological processes, 24(16): 1-15.

MWR, 1999. Abbay River Basin Master Plan Project., Minisry of Water Resource, Addis Ababa, Ethiopia.

Naqshband, S., Ribberink, J.S. and S.J.M.H., H., 2014. Using Both Free Surface Effect and Sediment Transport Mode Parameters in Defining the Morphology of River Dunes and Their Evolution to Upper Stage Plane Beds. ASCE, Journal of Hydraulic Engineering: 1-6.

NBI, 2003. Nile transboundary environmental action project. Project appraisal document, Nile Basin Initiative, Addis Ababa, Ethiopia.

NBI, 2004. Efficient water use for agricultural production, Nile Basin Initiative, Addis Ababa, Ethiopia.

Neitsch, S.L., Arnold, J.G., Kiniry, J. and Williams, J.R., 2011. Soil and water assessment tool theoretical documentation (Version 2009), Texas Water resources Institute. Texas AgriLife Research and USDA Agriculural Research Service, Temple, Texas, USA.

Nyssen, J., Poesen, J., Moeyersons, J., Haile, M. and Deckers, J., 2008. Dynamics of soil erosion rates and controlling factors in the Northern Ethiopian Highlands–towards a sediment budget. Earth Surface Processes and Landforms, 33(5): 695-711.

Omer, A.Y.A., 2011. Sedimentation of Sand and Silt in Roseires Reservoir: Vertical and Horizontal Sorting with Time, UNESCO-IHE, Delft, the Netherlands (Msc thesis).

Ottner, F., Gier, S., Kuderna, M. and Schwaighofer, B., 2000. Results of an inter-laboratory comparison of methods for quantitative clay analysis. Applied Clay Science, 17(5-6): 223-243.

Paarlberg, A.J., Dohmen-Janssen, C.M., Hulscher, S.J.M.H. and Termes, P., 2007. A parameterization of flow separation over subaqueous dunes. Water Resources Research, 43: W12417.

Paarlberg, A.J., Dohmen-Janssen, C.M., Hulscher, S.J.M.H., Termes, P. and Schielen, R., 2010. Modelling the effect of time-dependent river dune evolution on bed roughness and stage. Earth Surface Processes and Landforms, 35(15): 1854-1866.

Partheniades, E., 1962. A study of erosion and deposition of cohesive soils in salt water, University of California Berkeley, California (PhD thesis).

Partheniades, E., 1964. A summary of the present knowledge of the behavior of fine sediments in estuaries. Hydrodynamics Lab., Department of Civil and Sanitary Engineering, Massachusetts Inst. of Technology.

Peggy, A.J. and Curtis, D., 1994. Water Balance of the Blue Nile Basin in Ethiopia. Journal of Irrigation and Drainage Engineering, 120(3): 573-590.

Prinsen, G.F. and Becker, B.P.J., 2011. Application of SOBEK hydraulic surface water models in the Netherlands Hydrological Modelling Instrument. Irrigation and Drainage 60(Suppl. 1): 35-41.

Rădoane, M. and Rădoane, N., 2005. Dams, sediment sources and reservoir silting in Romania. Geomorphology, 71(1-2): 112-125.

Rimstidt, J.D., 1997. Quartz solubility at low temperatures. Geochimica et Cosmochimica Acta, 61(13): 2553-2558.

Roelvink, J.A., 2006. Coastal morphodynamic evolution techniques. Journal of Coastal Engineering, 53: 177-187.

Rosgen, D.L., 1994. A classification of natural rivers. Catena, 22(3): 169-199.

Rostamian, R., Jaleh, A., Heidarpour, M., Jalalian, A. and Abbaspour, K.C., 2008. Application of a SWAT model for estimating runoff and sediment in two mountainous basins in central Iran. Hydrological Sciences Journal, 53(5): 977-988.

Rouholahnejad, E., Abbaspour, K.C., Vejdani, M., Srinivasan, R., Schulin, R. and Lehmann, A., 2012. A parallelization framework for calibration of hydrological models. Environmental Modelling and Software, 31: 28-36.

Ruan, C.D. and Ward, C.R., 2002. Quantitative X-ray powder diffraction analysis of clay minerals in Australian coals using Rietveld methods. Applied Clay Science, 21(5-6): 227-240.

Salini and SP, 2010. 5000 Hydroelectric project, Ethiopian Electric Power Corporation, Ethiopia.

Santhi, C., Arnold, J.G., Williams, J.R., Srinivasan, R. and Hauck, L.M., 2001. Validation of the SWAT model on a large river basin with point and nonpoint sources. J. American Water Resources Assoc., 37(5): 1169-1188.

Schneiderman, E.M., Steenhuis, T.S., Thongs, D.J., Easton, Z.M., Zion, M.S., Neal, A.L., Mendoza, G.F. and Todd Walter, M., 2007. Incorporating variable source area hydrology into a curve-number-based watershed model. Hydrological Processes, 21(25): 3420-3430.

SCS, 1972. Hydrology. In: V. Mockus (Ed.), National Engineering Handbook, USDA Soil Conservation Service, USA.

Seleshi, Y. and Zanke, U., 2004. Recent changes in rainfall and rainy days in Ethiopia. International Journal of Climatology, 24(8): 973-983.

Setegn, S.G., Rayner, D., Melesse, A.M., Dargahi, B. and Srinivasan, R., 2011. Impact of climate change on the hydroclimatology of Lake Tana Basin, Ethiopia. Water Resources Research, 47(4).

Setegn, S.G., Srinivasan, R., Dargahi, B. and Melesse, A.M., 2009. Spatial delineation of soil erosion vulnerability in the Lake Tana Basin, Ethiopia. Hydrological Processes, 23(26): 3738-3750.

Shahin, M., 1985. Hydrology of the Nile Basin, Elsevier Science Publishers B.V./Science and Technology Division, P.O.Box 330,1000 A H Amesterdam, the Netherlands.

Sheridan, G.J., Lane, P.N.J., Sherwin, C.B. and Noske, P.J., 2011. Post-fire changes in sediment rating curves in a wet Eucalyptus forest in SE Australia. Journal of Hydrology, 409(1-2): 183-195.

Sima, B.A., 2011. Flow Regime and Land Cover Changes In the Didessa Sub basin of the Blue Nile River, South - Western Ethiopia, Swedish University of Agricultural Sciences (SLU), Uppsala, Sweden.

Singh, J., Knapp, H.V. and Demissie, M., 2004. Hydrologic modeling of the Iroquois River watershed using HSPF and SWAT, ISWS CR 2004-08. Champaign, Ill.: Illinois State Water Survey.

Siyam, A.M., El zein, S., sayed.S.M, E., Mirghani, M. and Golla, S., 2005. Assessment of the current state of the Nile Basin Reservoirs Sedimentation Problems (Group 1). NBCBN-River Morphology Research Cluster.

Sloff, K.J., 2007. SOBEK-RE exercises Handout. UNESCO-IHE lecture note.

SMEC, 2012. Roseires Dam Heightening Project, Reservoir operation study, Ministry of Water Resources and Electricity, Dams Implemetation Unit, Khartoum, Sudan.

Smith, J.D. and McLean, S.R., 1977. Boundary layer adjustments to bottom topography and suspended sediment, The 8th International Liege Colloqium on Ocean Dynamics. Elsevier Oceanography Series, pp. 123-151.

Sneath, P.H.A. and Sokal, R.R., 1973. SneNumerical taxonomy: the principles and practice of numerical classification, San Francisco: Freeman.

Stanley, D.J., 1996. Nile delta: extreme case of sediment entrapment on a delta plain and consequent coastal land loss. Marine Geology, 129: 189-195.

Steenhuis, T.S., Collick, A.S., Easton, Z.M., Adgo, E. and Ahmed, A.A., 2009. Predicting discharge and sediment for the Abay (Blue Nile) with a simple model. Hydrological Processes, 23(26): 3728-3737.

Stehr, A., Debels, P., Romero, F. and Alcayaga, H., 2008. Hydrological modelling with SWAT under conditions of limited data availability: evaluation of results from a Chilean case study. Hydrological Sciences Journal, 53(3): 588-601.

Sutcliffe, J.V., 2009. The hydrology of the Nile basin. In: H.J. Dumont (Ed.), The Nile Origin, Environments, Limnology and Human Use. Monographiae Biologicae, pp. 335-364.

Sutcliffe, J.V. and Parks, Y.P., 1999. The Hydrology of the Nile. IAHS Specific Publications no. 5.

Taj Elsir, A.M., Siddig, E.A. and Osman, E.H., 2001. Sediment Transport Management Practices in Reservoirs, Irrigation Canals and Pumps Schemes Inlets. FRIEND/NILE PROJECT, Proceedings of Sediment Transport and watershed Management Workshop, Khartoum-Sudan.

Talbot, M.R. and Williams, M.A.J., 2009. Cenozoic evolution of the Nile basin. In: H.J. Dumont (Ed.), The Nile Origin, Environments, Limnology and Human Use. Monographiae Biologicae, pp. 37-60.

Tamene, L., Park, S.J., Dikau, R. and Vlek, P.L.G., 2006. Analysis of factors determining sediment yield variability in the highlands of northern Ethiopia. Geomorphology 76(2006): 76-91.

Teferi, E., Bewket, W., Uhlenbrook, S. and Wenninger, J., 2013. Understanding recent land use and land cover dynamics in the source region of the Upper Blue Nile, Ethiopia: Spatially explicit statistical modeling of systematic transitions. Agriculture, Ecosystems & Environment, 165(0): 98-117.

Tekle, K. and Hedlund, L., 2000. Land Cover Changes Between 1958 and 1986 in Kalu District, Southern Wello, Ethiopia. Mountain Research and Development, 20(1): 42-51.

Tekleab, S., Mohamed, Y. and Uhlenbrook, S., 2013. Hydro-climatic trends in the Abay/Upper Blue Nile basin, Ethiopia. Physics and Chemistry of the Earth, Parts A/B/C, 61–62(0): 32-42.

Tesemma, Z.K., Mohamed, Y.A. and Steenhuis, T.S., 2010. Trends in rainfall and runoff in the Blue Nile Basin: 1964–2003. Hydrological Processes, 24(25): 3747-3758.

Tiesler, R., 2009. Roseires Dam Heightening Construction Surveys. Monaro Surveying Services, Australia, pp. 1 - 15.

Tou, J.T. and Gonzalez, R.C., 1974. Pattern Recognition Principles, Addison-Wesley, London.

UNESCO, 2004. National Water Development Report for Ethiopia, Ministry of Water Resources,Addis Ababa, Ethipoia.

van der Zaag, P. and Belay, S., 2007. In Search of Sustainable Catchments and Basin-wide Solidarities; Transboundary Water Management of the Blue Nile River Basin, WOTRO - Integrated Programme. .

van Griensven, A., Francos, A. and Bauwens, W., 2002. Sensitivity analysis and auto-calibration of an integral dynamic model for river water quality. Water Sci Technol, 45: 325-332.

van Griensven, A., Meixner, T., Grunwald, S., Bishop, T., Diluzio, M. and Srinivasan, R., 2006. A global sensitivity analysis tool for the parameters of multi-variable catchment models. Journal of Hydrology, 324(1–4): 10-23.

van Ledden, M., 2003. Sand-mud segregation in estuaries and tidal basins, Technical University of Delft, Delft, The Netherlands.

van Liew, M.W., Arnold, J.G. and Garbrecht, J.D., 2003. Hydrologic simulation on agricultural watersheds: Choosing between two models. Trans. ASAE, 46(6): 1539-1551.

van Rijn, L.C., 1984. Sediment transport, Part I: bed load transport. Journal of Hydraulic Engineering., ASCE, 129(10): 1613-1641.

Vazquez-Amábile, G.G. and Engel, B.A., 2005. Use of SWAT to compute groundwater table depth and streamflow in the Muscatatuck River watershed. Trans. ASAE 48(3): 991-1003.

Wagstaff, K., Cardie, C., Rogers, S. and Schrödl, S., 2001. Constrained k-means clustering with background knowledge, ICML, pp. 577-584.

Walling, D.E., 1977. Limitations of the rating curve technique for estimating suspended sediment loads, with particular reference to British rivers, In: Erosion and solid matter transport in inland waters (Proceedings of the Paris symposium, July 1977). IAHS Press Wallingford, UK pp. 34-38.

Walling, D.E., Collins, A.L. and Sichingabula, H.M., 2003. Using unsupported lead-210 measurements to investigate soil erosion and sediment delivery in a small Zambian catchment. Geomorphology, 52(3–4): 193-213.

Walling, D.E. and Webb, B.W., 1981. The reliability of suspended sediment load data, Erosion and Sediment Transport Measurement (Proceedings of the Florence Symposium, June 1981). IAHS Press Wallingford, UK pp. 177-194.

Wang, X., Chen, Q., Hu, H. and Yin, Z., 2011. Solubility and dissolution kinetics of quartz in NH_3–H_2O system at 25°C. Hydrometallurgy, 107(1–2): 22-28.

Waterbury, J., 1979. Hydropolitics of the Nile Valley. Syracuse University Press, New York.

White, E.D., Easton, Z.M., Fuka, D.R., Selassie, Y.G. and Steenhuis, T.S., 2011. Development and application of a physically based landscape water balance in the SWAT model. Hydrological Processes, 25(6): 915-925.

White, R., 2001. Review of sedimentation in reservoirs. In: T. Telford (Ed.), Evacuation of sediments from reservoirs, London, pp. 15-36.

Wilbers, A.W.E. and Ten Brinke, W.B.M., 2003. The response of subaqueous dunes to floods in sand and gravel bed reaches of the Dutch Rhine. Sedimentology, 50(6): 1013-1034.

Williams, J.R., 1995. The EPIC Model. In: V.P. Singh (Ed.), Computer Models of Watershed Hydrology. Water Resources Publications, Colorado, USA, pp. 909-1000.

Williams, M.A.J., 2009. Late Pleistocene and Holocene environments in the Nile basin. Global and Planetary Change, 69(1–2): 1-15.

Williams, M.A.J. and Talbot, M.R., 2009. Late quaternary environments in the Nile Basin. In: H.J. Dumont (Ed.), The Nile Origin, Environments, Limnology and Human Use. Monographiae Biologicae, pp. 61-72.

Winchell, M., Srinivasan, R., Di Luzio, M. and Arnold, J.G., 2010. ARCSWAT interface for Soil and Water Assessment Tool (SWAT 2009) - User's Guide, Texas AgriLife Research and USDA Agriculural Research Service, Temple,Texas- USA.

Wolela, A., 2007. Source rock potentialof the Blue Nile (Abay) Basin, Ethiopia. Journal of Petroleum Geology, 30(4): 389-402.

Wolela, A., 2008. Sedimentation of the Triassic–Jurassic Adigrat Sandstone Formation, Blue Nile (Abay) Basin, Ethiopia. Journal of African Earth Sciences, 52(1–2): 30-42.

Xie, Q., Chen, T., Zhou, H., Lu, H. and Balsam, W., 2013. Mechanism of palygorskite formation in the Red Clay Formation on the Chinese Loess Plateau, northwest China. Geoderma, 192: 39-49.

Yang, C.T., Huang, J. and Greimann, B.P., 2004-2005. *User's manual for GSTAR-ID (Generalized Sediment Transport model for Alluvial Rivers - One Dimension),version 1.0*. U.S. Bureau of Reclamation, Technical Service Center, Denver, Colorado.

Yang, C.T. and Simaes, F.J.M., 2000. User's Manual for GSTARS 2.1 (Generalized Sediment Transport model for Alluvial River Simulation version 2.1). Bureau of Reclamation, Technical Service Center, Denver, Colorado.

Yang, C.T. and Simaes, F.J.M., 2002. *User's Manual for GSTARS3 (Generalized Sediment Transport Model for Alluvial River Simulation version 3.0)*. Bureau of Reclamation, Technical Service Center, Denver, Colorado.

Yang, G., Chen, Z., Wang, Z., Zhao, Y. and Wang, Z., 2007. Sediment rating parameters and their implications: Yangtze River, China. Geomorphology, 85(3–4): 166-175.

Yates, D. and Strzepek, K., 1998 b. Modeling the Nile basin under climate change. Journal of Hydrologic Engineering, 3(2): 98-108.

Yevdjevich, V.M., 1965. Stochastic problems in design of reservoirs. In: C.S. University (Ed.), Paper presented at the seminar in Water Resources Research. , Colorado State , USA.

Yuan, W., Yin, D., Finlayson, B. and Chen, Z., 2012. Assessing the potential for change in the middle Yangtze River channel following impoundment of the Three Gorges Dam. Geomorphology, 147–148(0): 27-34.

Zeleke, G. and Hurni, H., 2001. Implications of landuse and land cover dynamics for mountain resource degradation in the northwestern Ethiopian highlands. Mount. Res. Develop, 21: 184-191.

Zhang, W., Wei, X., Jinhai, Z., Yuliang, Z. and Zhang, Y., 2012. Estimating suspended sediment loads in the Pearl River Delta region using sediment rating curves. Continental Shelf Research, 38(0): 35-46.

Samenvatting

De snelle bevolkingstoename en de toename van akkerland ten koste van bossen heeft geleid tot meer erosie in het bovenstroomse deel van het stroomgebied van de Blauwe Nijl. Bodemerosie vermindert de vruchtbaarheid van de grond en daarmee de opbrengst van de landbouw. Het geërodeerde materiaal komt terecht in de benedenstroomse Blauwe Nijl, waar op veel plekken sedimentatie plaatsvindt: in reservoirs van dammen leidt dit tot een vermindering van de bergingscapaciteit, wat problemen bij het opwekken van waterkracht tot gevolg heeft; in irrigatiekanalen leidt het sediment tot watertekorten en waterbeheersproblemen; in de rivier leidt sedimentatie to een verhoogd rivierbed en een verhoogd risico op overstromingen.

Momenteel vinden er in het stroomgebied van de Blauwe Nijl nog andere grote ontwikkelingen plaats, zowel in Ethiopië als Soedan. Ongeveer 30 kilometer bovenstrooms van de Ethiopisch-Soedanese grens wordt de Grand Ethiopian Renaissance Dam (GERD) gebouwd. Bovendien is 110 km benedenstrooms van de Ethiopisch-Soedanese grens recentelijk de Roseires dam met 10 meter verhoogd, waardoor 3700 miljoen m3 extra water geborgen kan worden. Plannen voor meer dammen in Ethiopië zijn in ontwikkeling om meer stroom op te wekken. Deze ontwikkelingen zullen de water- en sedimentstromen in de benedenstroomse Blauwe Nijl sterk veranderen.

Sedimentatie zal de levensduur van nieuwe reservoirs beïnvloeden, terwijl de hoeveelheid sediment die zal opslibben in irrigatiekanalen afhankelijk is van het beheer van de nieuwe dammen. De enige effectieve oplossing om de sedimentatieproblemen te verminderen is door de instroom van sediment naar de rivier te verminderen. Dit kan door maatregelen tegen erosie te nemen in het bovenstroomse deel van het stroomgebied. Gegeven de grote omvang van het bovenstroomse gebied is het belangrijk om te identificeren welke gebieden de grootste bronnen van sediment vormen.

Het hoofddoel van dit onderzoek is om deze gebieden te identificeren en om de hoeveelheden sediment uit deze gebieden te kwalificeren. Verder bestudeert dit onderzoek de effecten van nieuwe dammen op het sedimentatieproces.

In eerdere studies verzamelde data zijn ontoereikend om bovenstaande doelen te verwezenlijken. Het ontbreekt met name aan:

> ➤ bathymetrische en morfologische data, waaronder dwarsprofielen, en bodemmonsters van het rivierbed en de oevers van de rivier op verschillende plekken. Door moeilijke toegang tot de rivier is dit een grote lacune in de huidige kennis over de Blauwe Nijl.
> ➤ hydrologische data, waaronder rivierafvoer en sedimentconcentratie op verschillende plekken

> - de verdeling van water en sediment fluxen in het riviernetwerk bij verschillende riviercondities.
> - de geschiedenis van sedimentatie (waaronder de jaarlijkse hoeveelheden) in het benedenstroomse deel van het stroomgebied.
> - de oorsprong van het opgeslibte sediment in het benedestroomse deel van het stroomgebied.
> - de relatie tussen de verandering in landgebruik en de hoeveelheid sediment in de substroomgebieden, gegeven dat de grootste hoeveelheid sediment opslibt in het benedenstroomse deel van het stroomgebied.

Om deze lacunes te vullen is als onderdeel van dit onderzoek een uitbreide grensoverschrijdende meetcampagne uitgevoerd in Ethiopië en Soedan. De topografie van de rivierbedding is gemeten voor 26 dwarsprofielen van de Blauwe Nijl. In Ethiopië werd hier een ecosounder voor gebruikt en in Soedan een Acoustic Doppler Current Profiler. Er zijn bodemmonsters genomen van gebieden waar erosie plaatsvindt in het bovenstroomse deel van het stroomgebied en van de bedding en oevers van de rivier en zijtakken van de rivier. Op verschillende plekken in de rivier en in zijtakken in Ethiopië en Soedan zijn monsters van sediment in suspensie genomen. Gedurende de natte seizoenen in een periode van vier jaar is ook dagelijks de concentratie van het sediment in suspensie gemeten bij El-Deim station bij de Ethiopisch-Soedanese grens. De verzamelde monsters zijn geanalyseerd in Addis Ababa University Laboratory, het Hydraulics Research Station laboratory in Wad Medani en bij de Technische Universiteit Delft.

Jaarlijkse water- en sedimentbalansen werden verkregen door de beschikbare en gemeten afvoeren, de concentraties van sediment in suspensie te integreren in de ontwikkeling van een numeriek model: de Soil en Water Assessment Tool. De jaarlijkse sedimentbalansen werden benaderd voor verschillende plekken in de rivier en haar zijtakken. Drie regressiemethoden werden gebruikt om de sediment-gehalten te bepalen uit de rating-curves die werden vastgesteld op basis van de gemeten waarden. Deze methoden werden ontwikkeld met lineaire en non-lineare log-log regressies. Er is een statistische bias-correctiefactor toegepast om de lineaire regressieresultaten te verbeteren.

De waterbalans voor het hele riviersysteem werd onderzocht om de beschikbaarheid van water te bepalen voor ieder seizoen en voor alle riviercondities. Er is een eendimensionaal hydrodynamisch model van het gehele riviernetwerk ontwikkeld met Sobek software. In dit model zijn irrigatiewerken langs de rivier, alle belangrijke hydraulische kunstwerken en de beheersregels van deze kunstwerken opgenomen. Het model is verder gebruikt om het sediment transport te onderzoeken door het te integreren met de waterkwaliteitsmodule van Delft3D.

Dit geïntegreerde model (Sobek Rural, en Delfd3D Delwaq) maakte het mogelijk om de morfologische processen in de Blauwe Nijl te simuleren van het Tana-meer (het begin van de rivier) tot de Roseires dam. Het model is gekalibreerd en gevalideerd op basis van de historische bathymetrische onderzoeken van het Roseires Reservoir en de sedimentconcentraties gemeten op de Ethiopische -Sudanese grens. Vervolgens is de invloed van het verhogen van de Roseires Dam en de bouw van de Grand Ethiopian Renaissance Dam op de sedimentatiesnelheden gesimuleerd.

De sedimentatie-geschiedenis in Roseires Reservoir, de eerste plek waar sediment bezinkt in de Blauwe Nijl, werd bestudeerd door historische bathymetrische data te combineren met resultaten van een quasi-3D morfodynamisch model met verticale sortering (met Delft3D software). Selectieve sedimentatie zorgt voor bodemstratificatie in het reservoir, wat het mogelijk maakt om specifieke droge en natte jaren te herkennen. Het model maakte het mogelijk om gebieden in het reservoir aan te wijzen waar geen erosie en geen bank-migratie plaatsvond gedurende de levensduur van het reservoir. Deze informatie werd gebruikt om de meest veelbelovende bodemmonsterlocaties in het reservoir te bepalen. Een tweede meetcampagne werd uitgevoerd op deze locaties om het sediment in het reservoir te analyseren.

De oorsprong van het opgeslibte sediment in het benedenstroomse deel van het stroomgebied werd bepaald op basis van de mineraal-karakteristieken van het sediment. X-Ray Diffraction laboratoriumonderzoek maakte het mogelijk om de aanwezigheid van mineralen te bepalen in de sedimentmonsters genomen in de eroderende gebieden in het bovenstroomse deel van het stroomgebied en in het obgeslibte sediment in het Roseires Reservoir. De integratie van de resultaten van de X-Ray Diffractie door een cluster analyse maakte het mogelijk om de bron van de sedimenten in Roseires Reservoir te bepalen. De resultaten laten zien dat de substroomgebieden van Jemma, Didessa en Zuid Gojam de belangrijkste brongebieden voor sediment in de Blauwe Nijl zijn.

De implementatie van erosiereducerende maatregelen zou daarom in deze sub-stroomgebieden kunnen starten. In deze sub-stroomgebieden is het aandeel van natuurlijk bos, geplant bos, bosrijk grasland en grasland verminderd van meer dan 70% naar minder dan 25% van het oppervlak. Het gecultiveerde oppervlak nam toe van 30% tot meer dan 70%.

Tenslotte demonstreerden modelresultaten dat de jaarlijkse sedimentatie in de Grand Ethiopian Renaissance Dam (in aanbouw) zal variëren over de tijd, met maximum en minimum waarden van respectievelijk 45 en 17 miljoen kubieke meter sediment per jaar. In het Roseires Reservoir zal na de constructie van de Grand Ethiopian Renaissance Dam 2 miljoen kubieke meter sediment per jaar neerslaan. Dit betekent een reductie van meer dan 50% ten opzichte van de huidige situatie. Deze resultaten hebben een hoge onzekerheid, maar de trend en orde-grootte kunnen redelijk goed worden vastgesteld met het gekalibreerde en gevalideerde model.

ملخص

إن التزايد السكاني في أعالي حوض نهر النيل الأزرق أدي إلي تغيرات سريعة في استخدام الأراضي من غابات طبيعية إلي أراضي زراعية و التي بدورها تمخضت في تسارع عمليات تعرية التربة. فتعرية التربة تقلل من خصوبة التربة أعالي الحوض و بالتالي الإنتاجية الزراعية. فالتربة المنجرفة تنتقل إلي أدني حوض نهر النيل الأزرق، حيث تتم عملية الترسيب في أماكن متعددة. ففي بحيرات الخزانات، يؤدي الإطماء إلي تقليل السعة التخزينية، مما يخلق مشاكل في التوليد المائي كما يؤثر سلباً في النظم الإجتماعية۔ الإقتصادية و البيئية و الإحيائية. كما أن الطمي المترسب في قنوات الري يؤدي بدوره إلي نقصان و مشاكل في إدارة المياه. بالإضافة إلي ذلك فإن الطمي المترسب بقاع المجري الرئيسي للنهر يؤدي إلي إرتفاع منسوب القاع و الذي يعظم من مخاطر الفيضان.

إن حوض نهر النيل الأزرق يشهد في الوقت الراهن العديد من المنشآت التنموية في كل من السودان و أثيوبيا. مثال لذلك سد النهضة (الألفية) الأثيوبي ۔ تحت الإنشاء، و الذي يقع علي بعد 30 كلم تقريباً من الحدود الأثيوبية السودانية. كما تمت حديثاً تعلية خزان الروصيرص بالسودان، و الذي يقع علي بعد 110 كلم من الحدود، إلي 10 متر بغرض زيادة السعة التخزينية للبحيرة بحوالي 3,700 مليون متر مكعب إضافة لسعته التصميمية. إيضاً هنالك بعض السدود المخطط تشييدها في أثيوبيا بغرض التوليد المائي. كل هذه المنشآت التنموية سوف تؤثر علي الموارد المائية و معدلات الإطماء أدني حوض النيل الأزرق.

إن تشغيل بحيرات السدود الجديدة سوف لها دور فعال في معدلات الإطماء والتي سوف تؤثر علي العمر الإفتراضي لها و تبعاً لذلك كميات الطمي التي تترسب بقنوات الري. الحل الفعال لتقليل مشاكل الإطماء هو تقليل كميات الطمي الواردة من المصدر و هذا يتحقق بوسائل التحكم في جرف التربة في أعالي الحوض و يتطلب ذلك تحديد المواقع الأكثر قابلية للجرف بالحوض.

يهدف هذا البحث إلي تحديد المواقع الأكثر قابلية للجرف بالحوض و تقدير كميات الطمي الواردة من تلك المواقع. كما يتناول البحث دراسة آثار المنشآت التنموية بالحوض علي عمليات الإطماء.

هنالك العديد من الفجوات المعرفية و التي ينبغي تناولها لتحقيق الغايات المستهدفة من هذا البحث و التي تتعلق بالآتي:

➤ البيانات و المعلومات الهيدروغرافية و المورفولوجية المتمثلة في قطاعات النهر و خصائص تربة الضفاف و القاع و التي تعزي لصعوبة الوصول إلي مجري النهر في أعالي الحوض؛

➤ البيانات الهيدرولوجية مثل التصريفات و تركيز الطمي في العديد من المواقع؛

- توزيع التصريفات و حركة الطمي علي طول شبكة النهر في مختلف حالات التصريف؛
- البيانات و المعلومات التاريخية للإطماء علي مستوي الحوض؛
- مصدر الطمي المترسب في أدني الحوض؛
- العلاقات بين التغيرات في إستخدام الاراضي و إيرادات الطمي علي المستويات المحلية بالحوض و التي تجلب كميات كبيرة من الطمي المترسب في أدني الحوض.

لتقليل الفجوة المعرفية، فقد تمت أعمال مساحية مكثفة في كل من أثيوبيا و السودان كجزء من هذا البحث. تم قياس طبوغرافية القاع عند 26 قطاعاً عرضياً بالمجري الرئيسي للنهر و روافده و ذلك بإستخدام جهاز رجع الصدي (ecosounder) في أثيوبيا و جهاز قياس السرعات و الأعماق (ADCP) في السودان. أخذت عينات للتربة من المناطق المتأثرة بالجرف في أعالي الحوض كما أخذت عينات من قاع و ضفاف المجري الرئيسي للنهر و روافده. أخذت أيضاً عينات للطمي المتعلق من عدة مواقع علي إمتداد النهر و روافده في كل من أثيوبيا و السودان. بيانات الطمي المتعلق تم أخذها علي المستوي اليومي عند المنطقة الحدودية بين أثيوبيا و السودان في فترة الخريف و لمدة أربعة سنين. تم تحليل العينات المجمعة (التربة و الطمي) بمعامل جامعة أديس أبابا، محطة البحوث الهيدرولكية بالسودان و الجامعة التقنية دلفت.

تم عمل الموازنة للتصريفات و كميات الطمي السنوية عبر تكامل البيانات المتاحة و المجمعة حديثاً للتصريفات و تركيز الطمي و النماذج الحسابية مبنية علي برنامج آلية تقييم التربة و الماء (SWAT). الموازنة السنوية للطمي قدرت عند مواقع مختلفة بالمجري الرئيسي للنهر و روافده.

أستخدمت ثلاثة طرق لتحديد كميات الطمي بإستخدام منحنيات العلاقة بين التصريف و الطمي المقاسة. تتمثل هذه الطرق في العلاقة اللوغرثمية الخطية، اللوغرثمية الغير- خطية بينما تم إستخدام معامل التصحيح الإحصائي المحايد لتحسين نتائج العلاقات الخطية.

تم تقييم توزيع المياه علي طول مجري النهر لتحديد الموارد المائية المتاحة في كل المواسم و تحت مختلف الظروف. تم تطوير النموذج الهيدروديناكمي أحادي الأبعاد ليشمل كل إستخدامات المياه للري، بالإضافة إلي المنشآت الهيدرولكية الرئيسية و قوانين تشغيلها مستخدمين برنامج سوبك (SOBEK). حيث تم لاحقاً إستخدام نفس النموذج لدراسة حركة الطمي عبر التكامل مع جزئية برنامج دلفت 3د (DELFT3D) الخاص بنوعية المياه. إن التكامل بين برنامجي سوبك و دلفت3د يتيح محاكاة العمليات المورفولوجية بمنظومة نهر النيل الأزرق من منبعه ببحيرة تانا و حتي خزان الروصيرص. تمت معايرة و التحقيق للنموذج بواسطة البيانات التاريخية للمسح

الهيدروغرافي لبحيرة الروصيرص و تركيز الطمي المقاس عند الحدود الأثيوبية السودانية. ثم تم تشغيل النموذج للتنبؤ بأثر تعلية خزان الروصيرص و سد النهضة (تحت التشييد) علي معدلات الإطماء.

تمت دراسة تاريخ الإطماء داخل بحيرة خزان الروصيرص كأول مصيدة للطمي علي طول مجري النيل الأزرق، بتجميع البيانات التاريخية للمسح الهيدروغرافي مع نتائج النموذج المورفوديناميكي ثلاثي الأبعاد يحتوي الفرز الرأسي. طبقات التربة المترسبة داخل البحيرة تتيح التعرف علي السنوات الجافة و الرطبة. النموذج يتيح التعرف علي المناطق التي لم تتعرض للنحر الخالص أو الهدام منذ إنشاء الخزان. إن أكثر المواقع الواعدة لجمع عينات التربة، يمكن التعرف عليها عن طريق تحليل نتائج النموذج. أيضاً تمت قياسات ثانية بتلك المواقع لتحليل الطمي المترسب ببحيرة الخزان.ْ

تم التعرف علي أصل الطمي المترسب بالبحيرة عن طريق الخصائص المعدنية للمواد. ساهم التحليل بأشعة أكس (X-Ray) في تقييم المحتوي المعدني في عينات الطمي التي جمعت من مناطق التعرية في أعالي الحوض و طبقات التربة المترسبة ببحيرة خزان الروصيرص. التكامل بين نتائج التحليل بأشعة أكس و التحليل التجميعي، أدي إلي التعرف مصدر الطمي المترسب داخل بحيرة خزان الروصيرص. حيث أشارت النتائج إلي أن منطقة جمعة، ديدسسا و جنوب قوجام تمثل المناطق الرئيسية لمصدر الطمي المترسب.

عليه فإن ممارسات التحكم في النحر يمكن أن تبدأ من هذه المناطق حيث أن في الأربعين سنة الماضية، قد تناقص نسبة تغطية الغابات الطبيعية و الأراضي الخشبية و الرعوية من 70% إلي 25%. و بدلاً عن ذلك فقد تزايدت الأراضي الزراعية من 30% إلي أكثر من 70% من جملة المساحة الكلية.

أخيراً، أوضحت نتائج النموذج أن الإطماء المترسب سنوياً داخل سد النهضة سوف يتغير مع الزمن بقيم قصوي و دنيا 45 و 17 مليون متر مكعب/العام علي التوالي بمعدل ترسيب 27 مليون متر مكعب/العام. متوسط معدل الإطماء ببحيرة خزان الروصيرص بعد التعلية و تكملة تشييد سد النهضة سوف يكون 2 مليون متر مكعب/العام. هذا يعني أن معدلات الإطماء السنوي داخل بحيرة خزان الروصيرص سوف تتناقص بشكل ملحوظ بحوالي أكثر من 50% مقارنة مع الوضع الحالي. هذه النتائج تتأثر بمستوي علي من عدم التأكد و علي الرغم من ذلك يمكن أن نفترض أن نمط و مستوي القيم قد مثلت بصفة مناسبة بناءاً علي المعايرة و التحقيق للنموذج.

ABOUT THE AUTHOR

Yasir Salih Ahmed Ali, born on April 16, 1975 in Wad Medani, Sudan, obtained his bachelor degree with honor in Civil Engineering from the University of Khartoum in Sudan in 2000. Since 2001 he has been working at the Hydraulic Research Center, Ministry of Water Resources and Electricity in Sudan, conducting field work, analyzing data and preparing reports. He received his M.Sc in Water Management from the Water Management and Irrigation Institute, University of Gezira, Wad Medani, Sudan in 2006.

In April 2008, Yasir Salih Ahmed Ali obtained an M.Sc in Water Science and Engineering, specialization Hydraulic Engineering and River Basin Development, from UNESCO-IHE, Delft, the Netherlands.

From 2008 to 2014 he conducted his PhD study in Water Science and Engineering, Hydraulic Engineering and River Basin Development, UNESCO-IHE, Delft, the Netherlands.

Peer-reviewed international journal publications

- **ALI Y.S.A.**, CROSATO A., MOHAMED Y.A., ABDALLA S.H., WRIGHT N.G., ROELVINK J.A. (2014). Water resource assessment along Blue Nile River with a one-dimensional model. Proceedings of ICE Water Management, Vol. 167, No 7, pp. 394-413. DOI: 10.1680/wama.13.00020.

- **ALI Y.S.A.**, CROSATO A., MOHAMED Y.A., ABDALLA S.H. & WRIGHT N.G. (2014). Sediment balances in the Blue Nile River Basin. International Journal of Sediment Research, Vol. 29, No 3, pp. 316-328.

- OMER A.Y.A., **ALI Y.S.A.**, CROSATO A., PARON P. ROELVINK J.A. & DASTGHEIB A. (2014). Modelling of sedimentation processes inside Roseires Reservoir (Sudan). Earth Surf. Dynam. Discuss., 2, 153-179, www.earth-surf-dynam-discuss.net/2/153/2014/. DOI:10.5194/esurfd-2-153-2014.

Book chapters

- CROSATO A., **ALI Y.S.A.** & PARON P. (accepted). *Nile River – Geomorphology and sediment issues*. In: The Status and Future of the World's Large Rivers, editors Helmut Habersack and Des Walling, Wiley.

Publications in conference proceedings

- **ALI Y.S.A.**, A. CROSATO (2009). *Morphological processes of Blue Nile River between Roseires and Sennar Dams.* In: the International Conference on Water Conservation in Arid Regions in King Abdulaziz University, Water Research Center, Jaddah- Kingdom of Saudi Arabia.

- **ALI Y.S.A**, J.A. ROELVINK, N.G. WRIGHT, Y.A. MOHAMED, A. CROSATO (2012). *Quantification of water uses along the Blue Nile River network using a one dimension (1D) hydrodynamic model.* In: NCR-days 2012, Book of Abstracts, R. Schielen (Ed.), Publ. of the Netherlands Centre for Riverstudies (NCR) 36- 2012, pp. 22-24. http://www.ncr-web.org/ncr-days/ncr-book-of-abstracts

- **ALI Y.S.A.**, A. CROSATO (2013). *Sediment balances in the Blue Nile River Basin.* In: NCR-Days 2013, Book of Abstracts, A. Crosato, 2013 (Ed.), Publ. of the Netherlands Centre for Riverstudies (NCR) 37-2013, pp. 3-23 - 3-24. http://www.ncr-web.org/ncr-days/ncr-book-of-abstracts

- **ALI Y.S.A.**, A. CROSATO, Y.A. MOHAMED, S.H. ABDALLA, N.G. WRIGHT, J.A. ROELVINK (2013). *Morphodynamics impacts of the Grand Renaissance Dam construction and Roseires Dam heightening along Blue Nile River.* In: The New Nile Perspectives conference, Khartoum, Sudan, May 2013.

- P. PARON, **Y.S.A. ALI**, A. CROSATO, Y.A. MOHAMED (2013). *Sediment budget and fingerprinting in the Blue Nile Basin:* Roseires Reservoir case study. In: The New Nile Perspectives conference, Khartoum, Sudan, May 2013.

- **ALI Y.S.A.**, A.Y.A. OMER AND A. CROSATO (2013). *Modelling of sedimentation processes inside Roseires Reservoir (Sudan).* In: The 8th Symposium on River, Coastal and Estuarine Morphodynamics (RCEM 2013), University of Cantabria ,Santander, Spain, June 2013.

- KHALID BIRO, IGBAL SALAH, AGEEL BUSHARA, MOHAMED ALI AND **YASIR S. A. ALI** (2014). *The Impact of Climate Change on Crop Productivity in the Eastern Nile Basin of Sudan.* In: Nile Basin Development Forum, Nirobi, Kenya, October 6-7.

- **YASIR S. A. ALI**, AMGAD Y. A. OMER, ALESSANDRA CROSATO, YASIR A. MOHAMED, PAOLO PARON, NIGEL G. WRIGHT (2014). *Linking sediment source to sink. Case study: the trans-boundary Blue Nile River.* In: Nile Basin Development Forum, Nirobi, Kenya, October 6-7.

- **YASIR S. A. ALI**, AHMED S. HAYATY AND YASIR A. MOHAMED (2014). *Modifying the Operation Rules of Jebel Aulia Reservoir for higher reservoir levels.* In: Nile Basin Development Forum, Nirobi, Kenya, October 6-7.

Printed and bound by CPI Group (UK) Ltd, Croydon, CR0 4YY

21/10/2024

01777096-0014